T0140187

Studies in Systems, Decision and Control

Volume 242

Series Editor

Janusz Kacprzyk, Systems Research Institute, Polish Academy of Sciences, Warsaw, Poland

The series "Studies in Systems, Decision and Control" (SSDC) covers both new developments and advances, as well as the state of the art, in the various areas of broadly perceived systems, decision making and control–quickly, up to date and with a high quality. The intent is to cover the theory, applications, and perspectives on the state of the art and future developments relevant to systems, decision making, control, complex processes and related areas, as embedded in the fields of engineering, computer science, physics, economics, social and life sciences, as well as the paradigms and methodologies behind them. The series contains monographs, textbooks, lecture notes and edited volumes in systems, decision making and control spanning the areas of Cyber-Physical Systems, Autonomous Systems, Sensor Networks, Control Systems, Energy Systems, Automotive Systems, Biological Systems, Vehicular Networking and Connected Vehicles, Aerospace Systems, Automation, Manufacturing, Smart Grids, Nonlinear Systems, Power Systems, Robotics, Social Systems, Economic Systems and other. Of particular value to both the contributors and the readership are the short publication timeframe and the world-wide distribution and exposure which enable both a wide and rapid dissemination of research output.

** Indexing: The books of this series are submitted to ISI, SCOPUS, DBLP, Ulrichs, MathSciNet, Current Mathematical Publications, Mathematical Reviews, Zentralblatt Math: MetaPress and Springerlink.

More information about this series at http://www.springer.com/series/13304

Mohamed Elhoseny · Aboul Ella Hassanien
Editors

Emerging Technologies for Connected Internet of Vehicles and Intelligent Transportation System Networks

Emerging Technologies for Connected and Smart Vehicles

 Springer

Editors
Mohamed Elhoseny
Faculty of Computers and Information
Sciences
Mansoura University
Mansoura, Egypt

Aboul Ella Hassanien
Department of Information Technology
Faculty of Computers and Information
Cairo University
Giza, Egypt

ISSN 2198-4182   ISSN 2198-4190   (electronic)
Studies in Systems, Decision and Control
ISBN 978-3-030-22775-3   ISBN 978-3-030-22773-9   (eBook)
https://doi.org/10.1007/978-3-030-22773-9

This Springer imprint is published by the registered company Springer Nature Switzerland AG
The registered company address is: Gewerbestrasse 11, 6330 Cham, Switzerland

Preface

Over the past decade, advances in vehicular communications and intelligent transportation systems (ITS) have been aimed at trimming down the fuel consumption by avoiding traffic congestion and enhancement of traffic safety while initiating a new application, i.e., mobile infotainment. To address the individual requirements of both safety and non-safety applications in the connected and smart vehicles field, there is a need to build up a new communication technology for the integrated solutions of vehicular communication and smart communications. The connected vehicles infrastructure can be of various models such as vehicle-to-vehicle (V2V), vehicle-to-infrastructure (V2I), and vehicle-to-everything (V2E). Due to the rapid growth in the connected vehicles, many research constraints need to be addressed, e.g., reliability and latency, appropriate scalable design of MAC and routing protocols, performance and adaptability to the changes in environment (node density and oscillation in network topology), and evaluation and validation of connected vehicles' protocols under the umbrella of coherent assumptions using simulation methodologies. In addition, the information shared among connected vehicles is of great importance and it is not yet clear what kind of privacy policies will be defined for the ITS networks. This book aims to emphasize the latest achievements to identify those aspects of connected vehicles and ITS networks that are identical to a traditional communication network in the broader spectrum. In this book, we present the concepts associated with vehicular communication systems and their applications in 11 chapters.

Mansoura, Egypt   Mohamed Elhoseny
Giza, Egypt   Aboul Ella Hassanien
July 2019

Contents

Energy Efficient Optimal Routing for Communication in VANETs via Clustering Model

Mohamed Elhoseny and K. Shankar

Abstract Vehicular Ad Hoc Network (VANET) is a kind of extraordinary remote ad hoc network, which has high node portability and quick topology changes. Clustering is a system for gathering nodes, making the network increasingly vigorous. With no consciousness of nodes, at some point, it comes up short on energy which causes execution issues in network and topology changes. At that point, it emerges a primary issue of energy in routing protocol, which endeavor node lifetime and link lifetime issues in the network. To broaden the energy effectiveness of V2V communication, proposed a clustering based optimization technique. This paper presents K-Medoid Clustering model to cluster the vehicle nodes and after that, energy efficient nodes are recognized for compelling communication. With the expectation of accomplishing energy efficient communication, efficient nodes are recognized from each cluster by a metaheuristic algorithm, for example, Enhanced Dragonfly Algorithm (EDA) which optimizes the parameter as minimum consumption of energy in VANET. The outcome exhibits that the V2V communication improves the energy efficiency in all vehicle nodes additionally it accomplishes less execution time contrasted with existing algorithms.

Keywords VANET · Vehicle nodes · Energy efficient routing · K-Medoid clustering · EDA and minimum energy consumption

M. Elhoseny (✉)
Faculty of Computers and Information, Mansoura University, Mansoura, Egypt
e-mail: Mohamed_elhoseny@mans.edu.eg

K. Shankar
School of Computing, Kalasalingam Academy of Research and Education, Anand Nagar, Krishnankoil 626126, Tamil Nadu, India
e-mail: shankarcrypto@gmail.com

© Springer Nature Switzerland AG 2020
M. Elhoseny and A. E. Hassanien (eds.), *Emerging Technologies for Connected Internet of Vehicles and Intelligent Transportation System Networks*, Studies in Systems, Decision and Control 242, https://doi.org/10.1007/978-3-030-22773-9_1

1   Introduction

Nowadays, roads can be viewed as huge frameworks which incorporate computers, sensors, road-side foundation components, and vehicles. VANETs give the stage to data trade between road clients and roadside foundation components without requiring any network supplier [1]. Many research endeavors that have explored different issues identified with V2I, V2V, and VRC territories on account of the vital job they are relied upon to play in Intelligent Transportation Systems (ITSs) [2]. With the exponential development of energy utilization in remote communications, Green communication has been drawing increasingly more consideration in recent year [3]. The technique set comprises of the dimension of transmission power and the sending probability. In each round, nodes pick the energy consumption to send traffic; at that point, nodes will refresh the sending likelihood dependent on the convictions [4]. Another protocol enhances energy efficiency and decreases the number of dead nodes in large-scale Wireless Sensor Networks (WSNs). The algorithm is to locate the base inactivity and energy efficient way in a lossy network [4, 5]. The energy saving protocol attempts to diminish the energy utilization of the network in the WSN and accordingly increment the operational lifetime, which additionally, as a rule, leads to the use of the shortest routing paths [6, 25]. The energy balancing protocol, then again, endeavors to adjust the energy utilization to anticipate partitioning of the network [7, 26]. Be that as it may, the VANET joins a dynamic topology with an outsized and variable network estimate, and, obviously, it's to help quick nature of vehicles [8, 27]. From the clustering and optimization model the optimal path through which data is transmitted [9] from a source node to a goal node and decide if to utilize the moving vehicles as a versatile transfer to transmit data dependent on the bearing of movement just as the areas of the source node and the destination node [8]. Indeed, different VANET ventures have been executed by different governments, businesses, and academic establishments around the globe in the most recent decade or so [1].

2   Review of Existing Research Papers

With the expansion in demand for information download among the vehicular clients, control utilization, both at vehicle end and the roadside unit is additionally expanding relatively by Shrivastava et al. [10]. An improved multicast based energy efficient artful information scheduling algorithm. We gauge optimum data rate and an optimum number of clients having great channel conditions, in this way deterring the need to know the channel state data at the transmitter. The most testing features in VANETs are their dynamic topology and versatility, where vehicles are moving at variable high speeds and in various directions. Conversely, the test in the WSN is in dealing with the constrained energy assets of the nodes, since the execution of WSNs emphatically relies upon their lifetime by Mohaisen et al. [11]. To conquer these difficulties, this exploration researches the impacts of various Quality of Service (QoS) parameters on forwarding decisions in an efficient distributed position based routing protocol and spotlights on data transmission estimation.

Despite that VANET is considered as a subclass of MANET, it has for disposition the high versatility of vehicles delivering the successive changes of network topology that include changing of the road and fluctuating node thickness of vehicles existing in this road by Samira Harrabi et al. in 2017 [12]. Limiting network overhead value, number of created clusters and had not considered the vehicles intrigues which characterized as any related information used to separate vehicle from another. In Sharma et al. [13], It's to assesses enhanced AODV course data based IEEE 802.11 g ad hoc network consolidating OFDM radio network interfaces along with DCF-MAC protocol by methods for OPNET Modeler TM to understand an energy-efficient IEEE 802.11 g network. A novel methodology of load adjusted routing is proposed to improve the network strength and battery lifetime in individual nodes by Agarwal et al. [14], Assuming variable energy dimensions of transmission in every vehicle, our examination builds up some upper limits on the partition of two back to back RSUs for almost load adjusted routing. The issue has been characterized for a straight network with uniform dissemination of vehicles more than the 1-D road.

2.1 The Significance of Energy Efficient Routing

The increase in traffic once a day is a major test for the general population of developing nations. Accordingly, the experts capable should concentrate on road security to make the road traffic as efficient as could reasonably be expected [17–20]. Because of IT advancement, the communication among vehicles over expansive spaces has coordinated the consideration of scientists towards efficient road traffic the executives [21]. Information broadcasting from such a large number of sources with the limitation of opportune and conveyance of message causes blockage issue in vehicular communication, in this way guaranteeing packet dropping, low energy efficiency and broadened delay [22–24]. The proposed research work has been committed to the optimal routing structure in an energy-efficient way by optimizing the energy consumption parameter.

3 System Model

This model we will analyze the Vehicular communication framework by energy efficient routing process, it's solved by the help of optimization. Essentially, build up the vehicle network topology for V1R communication or V2V communication model. Before finding the energy efficient routing, clustering and cluster head selection is performed, the detail examination of our inventive model intricately talked about in the underneath area.

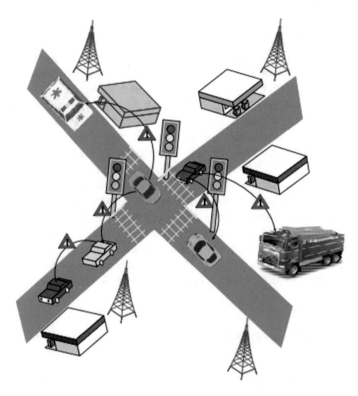

Fig. 1 Network topology

4   Network Topology

The network topology of networks can be viewed as a subset of the city map and the developments of vehicles are confined along the streets and by the traffic conditions. Here we have utilized 200 vehicle nodes, the ordinary nodes portrayed in gray color trade data among one another without GPS. Be that as it may, so as to have a decent knowledge of position data of the entire network, a couple of reference nodes are outfitted with GPS reliable and low-control communications are simultaneously considered and examined in light of the impedance of the jammers. The road separates the plane into two sections. The source node and destination node are on either side of the road, separately; the systematical model is exhibited in Fig. 1.

4.1   Vehicle Clustering

A few clustering strategies for VANET have been proposed in the literature. The cluster individuals and utilize the VANET clustering systems to frame the clusters, none of them took speed contrast into thought for clusters arrangement in VANET. Here we are utilized K-medoid clustering model to clusters the 200 vehicle nodes, from the group with fewer individuals are rejected and its cluster head joins the neighboring group, while different individuals begin group development process in the event that they can't join any close-by groups.

4.1.1 K-Medoid Clustering

This clustering model chooses "k" information things as beginning medoids to represent the k clusters. The various residual things are incorporated into a cluster which has its medoid nearest to them. From that point, another medoid is resolved which can represent the cluster better. K clusters are framed which are focused on the medoids and every one of the information individuals is put in the proper cluster dependent on closest medoid [15].

Initialization: Chooses K value of the n data points as the medoids

Medoids Selection: The medoid value is chosen by evaluating the distance between every two data points of all considered objects. Here, the distance measure is calculated by

$$\text{Dist} = \sum_{i=1}^{n} (F_i - S_i)^2 \tag{1}$$

Generation of Cluster: The initial cluster is formed by assigning each object to the closest medoid value.

Update the Selected Medoids: The role of calculating new medoid in each cluster is, it minimizes the total distance between objects in the cluster.

Vehicle Nodes into Each Medoid: By appointing each object to the closest medoid, the clustering result will be achieved. Assess the total sum of the separation from all objects for example instated n objects to their medoids. In the event that the calculated sum is equivalent to the past one, at that point end the execution of the clustering algorithm. Or else, rehash the method of K-medoid algorithm from the medoid determination step. By utilizing the K-medoid algorithm, 200 vehicle nodes are gathered dependent on their distance measure. The clustering process of V2V communication is represented in Fig. 2.

4.2 Head Selection

The determination criteria for the formation of cluster head mostly rely on the portability measurements (speed, position, and acceleration) of vehicle nodes.

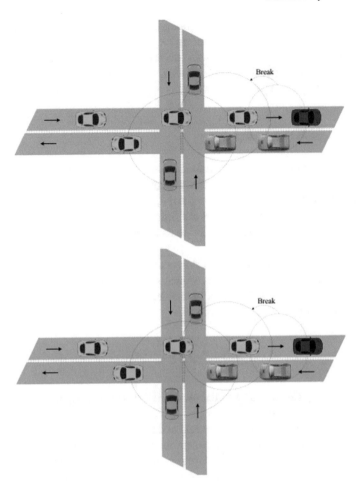

Fig. 2 Vehicle clustering

- A vehicle with the smallest average distance among a cluster is picked as the cluster head.
- *Closest Position to Average*: A vehicle endeavors to pick as its cluster head arranged by the absolute difference of candidate's position to the normal position of every single proximal vehicle.
- *Closest Velocity to Average*: A vehicle endeavors to pick as its cluster head arranged by the absolute difference of candidate's velocity to the average velocity of every single proximal vehicle.

4.3   Energy Efficient Optimization Model

We explore the optimal routing path configuration in traffic areas by considering the per-node most extreme energy effectiveness. Here, the energy efficiency between vehicular nodes is improved by the AODV protocol. The optimal energy efficient routing way is chosen by the parameter as minimum energy utilization.

4.3.1   Optimal Energy Efficient Route Selection

The primary objective is to build up a V2V communication with efficient routing by the network parameter optimization. The fitness work is determined dependent on the network parameter by minimizing the energy utilization of vehicular nodes. The target work is expressed as in condition (2).

$$OF(V2V) = \{\min(EC)\} \qquad (2)$$

The preferred minimized value is obtained by the use of inspired optimization technique i.e. Enhanced Dragonfly Algorithm (EDA).

Dragonfly Algorithm (DA): Dragonflies are considered as little predators that chase practically all other small insects in nature. The fundamental motivation of the DA algorithm starts with static and dynamic swarming practices [16]. These two swarming practices are fundamentally the same as the two principle periods of optimization utilizing meta-heuristics: exploration and exploitation.

Enhanced Function: The Cauchy mutation operator is acquainted with disregard to fall into a neighborhood ideal. Into a nearby optimum, this mutation operator is well ready to minimize the possibility of catching. The mutation probability for the worldwide best is equivalent to zero and it upgrades with diminishing the fitness by applying this new mutation probability.

Implementation Steps of EDA

(i) **Initialization:** Initialize the population of dragonflies (vehicle nodes) in terms of V_i

$$V_i = V_1, V_2, V_3, \ldots\ldots V_n, \text{ where } i = 1, 2, 3 \ldots n. \qquad (3)$$

(ii) **Behavior Analysis of DA**

The behavior of Dragonflies: The behavior of dragonflies can be explained in five steps, namely, separation, alignment, cohesion, attraction towards a food source and distraction outwards an enemy (Table 1).

Parameter Description: In separation, $Se_i \rightarrow$ indicates the separation of i-th individual, V is the current position of the individual, V_k is the position of k-th individual, N is the total number of neighboring individual in the search space. In alignment, $Al_i \rightarrow$ indicates the alignment of i-th neighboring individual, v_k is the

Table 1 The behavior of
dragonfly's calculation

Behavior of Dragonflies	Formula
Separation	$Se_i = \sum_{k=1}^{N} V - V_k$
Alignment	$Al_i = \frac{\sum_{k=1}^{N} v_k}{N}$
Cohesion	$Co_i = \frac{\sum_{k=1}^{N} V_k}{N} - V$
Attraction towards a food source	$Food_i = V^+ - V$
Distraction outwards an enemy	$Enemy_i = V^- + V$

velocity of k-th individual. In food source and enemy, V^- indicates the position of
the enemy, V^+ indicates the position of food source.

(iii) _Updation process_: To update the position of artificial dragonflies in an inquiry
space and reproduce their developments, two vectors are considered: step (ΔD) and
position (V). The progression vector closely resembles the speed vector in Particle Swarm Optimization (PSO), and the DA algorithm is created dependent on the
framework of the PSO algorithm. The progression vector demonstrates the course
of the development of the dragonflies and characterized as pursues:

$$\Delta V_{t+1} = (sSe_i + aAl_i + cCo_i + fFood_i + eEnemy_i) + w\Delta V_t \quad (4)$$

where s signifies the separation weight, a signifies the alignment weight, c is the
cohesion weight, f signifies the food factor, e signifies enemy factor, w signifies
inertia weight, t shows iteration count.

The position vector can be calculated as

$$V_{t+1} = V_t + \Delta V_{t+1} \quad (5)$$

During the enhancement procedure, different explorative and exploitative practices
can be accomplished. At the point, there is no neighboring solution; the situation
of dragonflies is refreshed by Cauchy mutation probability. Along these lines, the
position vectors V are determined as:

$$V_{t+1} = V_t + M_p V_t \quad (6)$$

The neighborhood area is expanded and at last, at the conclusive period of the optimization process, the swarm becomes just a single gathering. Food source and the
enemy are chosen from the best and the most exceedingly bad arrangements got
in the entire swarm at any moment. This leads the assembly towards the promising locales of pursuit space and in the meantime, it leads dissimilarity outward the
non-promising territories in inquiry space. The flow diagram of the EDA is shown
in Fig. 3.

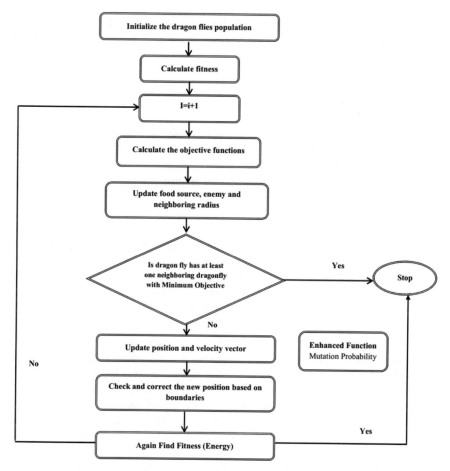

Fig. 3 Flowchart of EDA

In view of the achieved optimal parameter as minimized Energy utilization, the efficient vehicle nodes with its routing way are chosen. In course disclosure process, the trust module encourages the destination node to introduce the threshold value for the energy-factor and it additionally inspects the energy-factor of the nodes to incorporate or dismiss the nodes in the course.

5 Result and Discussion

In the result analysis part, the performance of the K-medoid clustering and optimization model is investigated by examining the network parameters in terms of energy consumption, packet delivery ratio, network lifetime and throughput. The execution of the proposed study is analyzed with the help of Network Simulation Tool-Version 2 (NS2). The simulation results of the proposed work are described in this section.

5.1 Evaluation Metrics

Throughput: Network throughput can be defined as the accomplishment of acknowledged packets over a communication channel.

Packet to the Delivery ratio (PDR): It can be described as a number of packets delivered successfully and correctly to the number of packets delivered over communication.

Energy Consumption: It is referred to as total energy depleted during data transmission of packets from source to destination.

Table 2 clarifies the outcome examination of a proposed model for an alternate number of vehicle nodes. In light of the node quality (efficient course), vehicles are clustered utilizing K-medoid clustering model. The execution of the proposed model is analyzed as far as throughput PDR, NLT, and EC. Contrasted with existing models, the energy efficient routing for V2V communication is practiced in K-medoid with EDA model.

In the throughput analysis, these parameters are analyzed dependent on the clustering and optimization procedures; showed in Fig. 4a. Compared to the customary Dragonfly algorithm, the enhanced Dragonfly algorithm gives the better throughput for 50, 100, 150 and 200 quantities of vehicles in the information transmission. The Packet to Delivery Ratio (PDR) for various quantities of the vehicle is exhibited in Fig. 4b. On contrasting the proposed model and the current methodology, the presented clustering K-medoid with EDA accomplishes high PDR for the 50, 100, 150 and 200 quantities of vehicles.

Figure 4 clarifies the execution of the vehicular network by estimating throughput, NLT, PDR, and EC individually under the diverse vehicle nodes. Figure 4c portrays the Network lifetime (days) for different numbers of a vehicle like 50, 100, 150 and 200 based optimization systems. At 50 quantities of vehicles, the k-medoid with EDA accomplishes the network lifetime as 8.5 days, k-medoid with EDA as 4 days. So also, for different quantities of the vehicle (100−200) the NLT is dissected and

Table 2 Performance Metrics Results

No. of Vehicles	Throughput (kbps)		PDR (%)		NLT (Day)		EC (J)	
	Existing	Proposed	Existing	Proposed	Existing	Proposed	Existing	Proposed
50	8524	10547	59.55	92.75	6.5	10	252	189
100	7284	8247	61.75	89.64	7	9.2	242	156
150	7124	8045	49.28	78.44	8.5	9.5	308	227
200	7025	8115	56.72	69.78	4.8	10.5	342	220

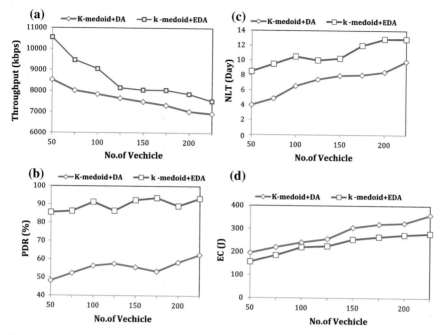

Fig. 4 Performance analysis **a** Throughput, **b** PDR, **c** NLT and **d** EC

compared. At long last, the diagram finishes up the proposed methodology achieves the most extreme lifetime in the V2V communication.

At whatever point the node transmission control is more, at that point, the number of bounces in between source to destination will be generally less. This thusly results in diminished control overhead and expanded feasible throughput. In any case, expanded node control utilization may lead to impedance with part nodes which results in packet loss because of crashes. So as to lessen the energy utilization of V2V communication between vehicle nodes, we proposed k-medoid clustering with EDA optimization. Figure 4d demonstrates the energy utilization rate of 50, 100, 150 and 200 vehicle nodes based communication. Contrasted with existing advancement, K-medoid clustering with EDA expends the least energy.

The percentage analysis of EC for a number of vehicle nodes is illustrated in Fig. 5. For 25 numbers of vehicles, k-medoid + DA consume 83% energy in data transmission and k-medoid + EDA consumes 49%. Similarly, 50, 75, 100, 125, 150, 175 and 200 numbers of vehicles are illustrated in the figure. The bar graph concludes that the proposed approach (k-medoid with EDA) achieves minimum energy consumption compared to k-medoid + DA approach.

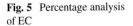

Fig. 5 Percentage analysis of EC

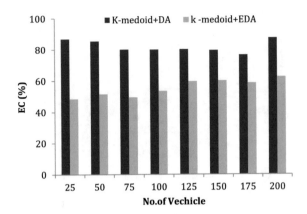

6 Conclusion

This paper introduced the efficient node identification of VANETs considering with energy utilization factor at least. This goal was accomplished by the clustering algorithm, for example, K-medoid algorithm along with optimization. It groups the vehicle nodes in various adjusts and choosing any nodes as cluster heads in certain rounds, it has the capacity to diminish the number of transmitted messages from every node to various nodes and to the base station, saving more energy in the network. At that point, the energy efficient route for V2V communication was obtained among the clustered nodes in VANETs by the advancement of network parameter (EC) utilizing the EDA algorithm. From the simulation examination, the proposed algorithm k-medoid + EDA give the minimum energy consumption, contrasted with the k-medoid + DA method. Later on, we can also broaden this work by improving energy efficiency regarding limited expense and maximized QoS by proposing progressively efficient clustering and optimization strategies.

References

1. Al-Mayouf, Y.R.B., Abdullah, N.F., Ismail, M., Al-Qaraawi, S.M., Mahdi, O.A., Khan, S.: Evaluation of efficient vehicular ad hoc networks based on a maximum distance routing algorithm. EURASIP Journal on Wireless Communications and Netw. **2016**(1), 265 (2016)
2. Rehman, O., Ould-Khaoua, M.: A hybrid relay node selection scheme for message dissemination in VANETs. Future Gener. Comput. Systems **93**, 1–17 (2019)
3. Huang, M., Yang, B., Ge, X., Xiang, W., Li, Q.: Reliable energy-efficient routing algorithm for vehicle-assisted wireless ad-hoc networks. In: 2018 14th International Wireless Communications & Mobile Computing Conference (IWCMC), pp. 1219–1224. IEEE (2018)
4. Lee, J.H., Moon, I.: Modeling and optimization of energy efficient routing in wireless sensor networks. Appl. Math. Model. **38**(7–8), 2280–2289 (2014)
5. Kakhandki, A.L., Hublikar, S.: Energy efficient selective hop selection optimization to maximize the lifetime of the wireless sensor network. Alex. Eng. J. **57**(2), 711–718 (2018)

6. Saiáns-Vázquez, J.V., López-Nores, M., Blanco-Fernández, Y., Ordóñez-Morales, E.F., Bravo-Torres, J.F., Pazos-Arias, J.J.: Efficient and viable intersection-based routing in VANETs on top of a virtualization layer. Ann. Telecommun. 1 –12 (2018)

7. Majumdar, S., Prasad, P.R., Kumar, S.S. Kumar, K.S.: An efficient routing algorithm based on ant colony optimization for VANETs. In: 2016 IEEE International Conference on Recent Trends in Electronics, Information & Communication Technology (RTEICT), pp. 436–440 (2016). IEEE

8. Hao, S., Zhang, H., Song, M.: A stable and energy-efficient routing algorithm based on learning automata theory for MANET. J. Commun. Inf. Netw. 3(2), 52–66 (2018)

9. Laroiya, Namita, Lekhi, Sushil: Energy efficient routing protocols in vanets. Adv. Comput. Sci. Technol. 10(5), 1371–1390 (2017)

10. Shrivastava, A., Bansod, P., Gupta, K., Merchant, S.N.: An improved multicast based energy efficient opportunistic data scheduling algorithm for VANET. AEU-Int. J. Electron. Commun. 83, 407–415 (2018)

11. Mohaisen, L.F., Joiner, L.L.: Interference-aware bandwidth estimation for load balancing in EMHR-energy based with mobility concerns hybrid routing protocol for VANET-WSN communication. Ad Hoc Netw. 66, 1–15 (2017)

12. Harrabi, S., Jaafar, I.B., Ghedira, K.: Message dissemination in vehicular networks on the basis of agent technology. Wirel. Pers. Commun. 96(4), 6129–6146 (2017)

13. Sharma, V., Singh, H., Kant, S.: AODV based energy efficient IEEE 802.16 G VANET network (2013)

14. Agarwal, S., Das, A., Das, N.: An efficient approach for load balancing in vehicular ad-hoc networks. In: 2016 IEEE International Conference on Advanced Networks and Telecommunications Systems (ANTS), pp. 1–6. IEEE (2016)

15. Mirjalili, S.: Dragonfly algorithm: a new meta-heuristic optimization technique for solving single-objective, discrete, and multi-objective problems. Neural Comput. Appl. 27(4), 1053–1073 (2016)

16. Yang, B., Zhang, Y.: Kernel-based K-medoids for clustering data with uncertainty. In: International Conference On Advanced Data Mining And Applications, pp. 246–253. Springer, Berlin (2010)

17. Murugan, B.S. Elhoseny, M., Shankar, K., Uthayakumar, J.: Region-based scalable smart system for anomaly detection in pedestrian walkways. Comput. Electr. Eng. 75, 146–160 (2019)

18. Shankar, K., Elhoseny, M., Chelvi, E.D., Lakshmanaprabu, S.K., Wu, W.: An efficient optimal key based chaos function for medical image security. IEEE Access 6, 77145–77154 (2018)

19. Elhoseny, M., Shankar, K., Lakshmanaprabu, S. K., Maseleno, A., Arunkumar, N.: Hybrid optimization with cryptography encryption for medical image security in Internet of Things. In: Neural Computing and Applications, pp. 1 –15. https://doi.org/10.1007/s00521-018-3801-x (2018)

20. Shankar, K., Elhoseny, M., Kumar, R. S., Lakshmanaprabu, S. K., Yuan, X.: Secret image sharing scheme with encrypted shadow images using optimal homomorphic encryption technique. J. Ambient Intell. Humanized Comput. 1 –13. https://doi.org/10.1007/s12652-018-1161-0(2018)

21. Gaber, T., Abdelwahab, S., Elhoseny, M., Hassanien, A.E.: Trust-based secure clustering in WSN-based intelligent transportation systems. Comput. Netw. https://doi.org/10.1016/j.comnet.2018.09.015 (2018). Accessed 17 Sept 2018

22. Mohamed, R.E., Ghanem, W.R., Khalil, A.T., Elhoseny, M., Sajjad, M., Mohamed, M.A.: Energy efficient collaborative proactive routing protocol for wireless sensor network. Comput. Netw. https://doi.org/10.1016/j.comnet.2018.06.010 (2018). Accessed 19 June 2018

23. Elhoseny, Mohamed, Tharwat, Alaa, Yuan, Xiaohui, Hassanien, A.E.: Optimizing K-coverage of mobile WSNs. Expert Syst. Appl. 92, 142–153 (2018)

24. Elsayed, Walaa, Elhoseny, Mohamed, Sabbeh, Sahar, Riad, Alaa: Self-maintenance model for wireless sensor networks. Comput. Electr. Eng. 70, 799–812 (2018)

25. Elhoseny, M., Tharwat, A., Farouk, A., Hassanien, A.E.: K-coverage model based on genetic algorithm to extend WSN lifetime. IEEE Sens. Lett. 1(4), 1–4 (2017). IEEE

26. Elhoseny, M., Farouk, A., Zhou, N., Wang, M.-M., Abdalla, S., Batle, J.: Dynamic multi-hop clustering in a wireless sensor network: performance improvement. Wirel. Pers. Commun. **95**(4), 3733–3753

27. Elhoseny, M., Yuan, X., Yu, Z., Mao, C., El-Minir, H., Riad, A.: Balancing energy consumption in heterogeneous wireless sensor networks using genetic algorithm. IEEE Commun. Lett. IEEE **19**(12), 2194–2197 (2015)

Mobility and QoS Analysis in VANET Using NMP with Salp Optimization Models

K. Shankar, M. Ilayaraja, K. Sathesh Kumar and Eswaran Perumal

Abstract Wireless Sensor Networks (WSN), the essential thing is to beat path selection and duration of VANET framework by most extreme mobility as well as transmission rate. In light of this mobility and connectivity charts changes in all respects as often as possible and it influences the execution of VANETs. Because of the attributes of VANET, for example, self-association, dynamic nature,and quick moving vehicles, routing in this network is an impressive test. This chapter discussed the mobility, just as Quality of Service (QoS) is VANET communication process, Here we are utilized Network Mobility Protocol (NMP) for routing reason, Moreover, the extensive optimization procedure is Salp Swarm Optimization (SA). It's able to improve the initial random solutions viably and merge towards the optimum, Based on optimization procedure to get an optimal path with a minimum delay from source to destination. From this NMP-SA model improve the high mobility of nodes. The simulation results demonstrate better QOS parameters contrasted with other similar optimization Models.

Keywords Vehicle ad hic network (VANET) · Mobility · Optimization · Swarm algorithm

K. Shankar (✉) · M. Ilayaraja · K. Sathesh Kumar
School of Computing, Kalasalingam Academy of Research and Education, Krishnankoil, India
e-mail: shankarcrypto@gmail.com

M. Ilayaraja
e-mail: ilayaraja.m@klu.ac.in

K. Sathesh Kumar
e-mail: sathesh.drl@gmail.com

E. Perumal
Department of Computer Applications, Alagappa University, Karaikudi, India
e-mail: eswaranperumal@gmail.com

© Springer Nature Switzerland AG 2020
M. Elhoseny and A. E. Hassanien (eds.), *Emerging Technologies for Connected Internet of Vehicles and Intelligent Transportation System Networks*, Studies in Systems, Decision and Control 242, https://doi.org/10.1007/978-3-030-22773-9_2

1 Introduction

In the ad hoc network, nodes are conveying straightforwardly to one another without utilizing any passage. The variable node thickness, high node mobility, and unusual and cruel communication condition [1]. Several routing protocols were presented to use network assets and improve routing effectiveness in VANETs [2]. Be that as it may, the vast majority of these protocols still have deficiencies when meeting the QoS [3] prerequisites and in ensuring the stability of network topology at the time of routing procedure [4]. In these circumstances, the most critical necessity is the transmission of data from source to objective with a specific dimension of QoS. The QoS implies the transmission of information from source to objective with minimum delay and minimum overhead [5]. Since portable vehicles are allowed to move arbitrarily, vehicle mobility is a standout amongst the most essential issues in protocol structure [6]. The impacts of vehicle mobility on traffic flow control, routing way determination, portable channel appointing, control overhead estimation. The routing is subject to the protocols being utilized for routing in the network [7, 18]. If there should be an occurrence of VANET, execution of these routing protocols relies on different situations like urban and the highway. The routing protocols which are position based are normally viewed as all around coordinated if there should be an occurrence of vehicular setting [8, 19–21]. VANETs present a difficult nature for protocol and request plan because of their low dormancy and raised information rate necessities in a raised mobility condition [9, 22].

For upgrading the mobility model the optimization will be utilized, optimizations methods like Salp Swam Optimization (SSO) evolutionary computation [8] and Particle Swarm Optimization (PSO) [9], and so on. Some of them are nearby inquiry-based and others (PSO, GA, ACO) are global search based heuristics [10]. Our proposed slap swarm model to get the best way with maximum number data transmission process over VANET Network [11].

2 Related Works

In 2015 Abdul Halim et al. [12], has been proposed the network optimization is one approach to keep up the current protocols and other network parameter contrast with structure and actualizing new improved protocols in which it excessively expensive. The need for optimization of the vehicular network is to achieve throughput, end-to-end delay and packet delivery ratio utilizing the Taguchi technique.

Grey wolf optimization based clustering calculation for VANETs is proposed, that imitates the social conduct and hunting system of grey wolfs for making effective clusters by Fahad et al. in 2018 [5]. The proposed technique is contrasted with well-known meta-heuristics from existing research papers and results demonstrate that it gives ideal results that lead to a vigorous routing protocol for grouping of VANETs, which is suitable for highways and can achieve quality communication, affirming dependable conveyance of data to every vehicle.

QoS-Aware Routing Protocol for VANETs called QoS-Aware Routing in VANETs (QARV) in which packets achieve the goal while satisfying the QoS (Kaur et al. 2018) [13]. This protocol works in the interstate situation. The new idea called Terminal Intersection idea is utilized in these two protocols so as to decrease the congestion and reduce the ideal opportunity for investigating the route. Two bio-roused calculations, Ant Colony Optimization (ACO) and Bee Colony Optimization (BCO), are utilized to accomplish the outcomes.

In 2013 Raw et al. [14], path duration can be utilized to anticipate the behavior of the versatile nodes in the network. Estimation of the path duration in VANETs can be a key factor to improve the execution of the routing protocol. Estimation of way length is a provoking assignment to execute as it relies upon numerous parameters including node density, transmission range, quantities of hops, and speed of nodes.

In the road, since each vehicle is moving a fixed way with high moving pace, the vehicle embracing our protocol can gain IP address from the VANET through the vehicle to vehicle communication by Chen et al. in 2009 [15]. The vehicle can depend on the help of the front vehicle to execute the pre-handoff system or it might gain its new IP address through multi-hop transfers from the vehicle on the paths of the equivalent or inverse course and in this manner lessens the handoff delay.

3 Mobility Analysis

Mobility models usually focus on the individual moving behavior between mobility ages. Here, an epoch is considered as brief timeframe, in which both moving speed and moving bearing of vehicles are roughly considered as consistent [23]. Mobility management is basic for high-speed and consistent administrations for vehicular networks since vehicle nodes change their purposes of connection much of the time and net-work topology can change suddenly. Mobility in VANET communication having a few necessities [17]:

Fast Vertical handover: This prerequisite is required for delay-sensitive applications, similar to wellbeing, to stay away from the accidents. In a heterogeneous wireless condition, vertical handover of the portable client's associations among various wireless innovations must be upheld to accomplish consistent administration [24].

Communication support: Data transmission range of the mobile nodes is to achieve the destination. Mobility the management plans for vehicular networks need to consider the multilevel communications necessities [25].

The efficiency of VANET: High frequency of change of the purpose of connection, the mobility management plot must be versatile and productive to help distinctive sorts of traffic [26].

Location Management: With novel mobility attributes of VANETs, fundamental routing protocols can't be legitimately connected to VANETs because of the large latency and overhead. The upgrade from geological clustering was utilized for the two areas refreshing and questioning to improve scalability. For territory awareness, neighborhood look was utilized to find the goal node [27].

3.1 Mathematical Expression

The mobile model examined with the accompanying suspicion that is the relative moving rate and the relative moving direction between two versatile vehicles are considered as two free irregular factors. At that point, the vehicle moving speed possibly haphazardly changes when inter-vehicle link changes. For this mobility, the termination time determined by:

$$ET(a, b) = \frac{\sum\limits_{a \in N(b)} ET}{|N(b)|} \qquad (1)$$

Condition (1) demonstrates the measure the capacity of a vehicle to keeping up availability with vehicles inside its transmission extend as the average lifetime of the connections built up with the neighboring vehicle. The vehicular average velocity difference is processed by

$$V(\text{diff}) = \sum\limits_{a \in N(b)} |\bar{b}(a) - \bar{a}(a)| / |N(a)| \qquad (2)$$

As it were, mobility models depict the example of the development of vehicles, and how their area, speed and acceleration change after some time. The execution of any protocol is predominantly controlled by the mobility patterns of vehicle, it is attractive to recreate the vehicle development of focused genuine applications in a sensible manner.

3.2 VANET Network Topology

Mobility management in VANET is essential for V2 V (vehicle to vehicle) communication. To upgrade the mobility and communication level between the Roadside unit and vehicle the optimization model is considered. This VANET structure will appear in Fig. 1, the source vehicles will secure the geographic spots of their different target location administrations and the absolutely unique transmission sets have a similar Quality of service that is most extreme network life time, throughput, and delivery proportion. A fixed node is put on each intersection to help packet diverting and routing information storage.

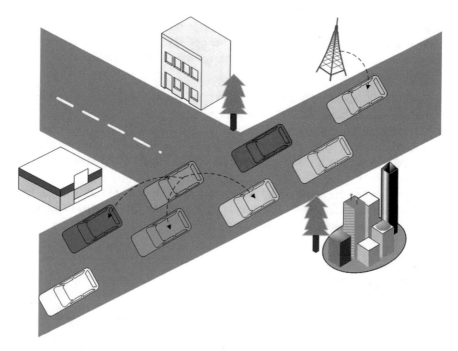

Fig. 1 VANET communication model

Sensitive data are exchanged between the optimal route nodes. These nodes can communicate with one another through different ways utilizing direct connections within the transmission range or utilizing middle nodes to exchange data packet from source to goal, the mobility and thickness of nodes changes with time and area. For this routing model imaginative protocol will be utilized, the explanation of this protocol is discussed in underneath segment.

3.2.1 Protocol for M-VANET

Our epic work, Network Mobility Protocol (NMP) utilized, it's an idea and the property of multi-hop VANETs. Every vehicle is outfitted with a mobile router, which is furnished with two communication interfaces, including the Wi-MAX and Wi-Fi interfaces. When it scopes to goal, it stays static for the predefined stop time and moving again as indicated by the same principle. The mobility behavior of nodes especially relies upon the stops time and most extreme speed of nodes. This mobility protocol the procedure determined by

$$m(t) = m(t-1) + \delta * \text{acce}(t) \tag{3}$$

Mobility model, the acceleration is input, to ascertain nodes next speed the acceleration of Current node. δ It implies that the node is moving with de-acceleration (negative increasing speed) Otherwise, it moves with positive acceleration.

Pros of NMP

- Router Free Connection to the web with no wireless switch is the primary preferred standpoint of utilizing a mobile ad hoc network.
- The control of the network is disseminated among the mobile nodes of the network as there is no foundation network for the principle control of the network activity.
- It lessens inordinate memory prerequisites and the NMP decreases the route repetition.
- This protocol is the constraining the number of encapsulation and passages two along these lines diminishing the packet delivery delay. The number of encapsulation and passages in intra-NMP routing is zero.

3.3 Proposed Architecture

A new route or path must be set up for further communication once a path failure occurs. Expanding the proficiency of execution of VANETs learning of the path duration can help extraordinarily. For the optimal route determination swarm based optimization used, the purpose behind choosing the optimal path is to improve mobility and QoS factors. Notwithstanding the distinction between developmental and swarm knowledge systems, the basic is the improvement of one or a lot of solutions at the process of optimization. The speed and the direction of the node determined by the

$$\text{Speed}_n = \beta * s_{p-1} + (1 - \beta) * s + (1 - \beta^2) * s_{dn-1^{1/2}} \qquad (4)$$

Current speed and direction of the mobile node are dependent on the past speed and direction values. In view of the speed vehicle mobility in a simulation considers the probability appropriation capacities covering both moving direction and speed consistently as the recreation advances. The estimations of speed and direction for development in the period; β is the consistent incentive in the range [0, 1].

3.4 Optimization Process

Salps have a place with the group of Salpidae and have a straightforward barrel-formed body. Their tissues are exceedingly like jelly fishes. They likewise move fundamentally the same as jellyfish, in which the water is pumped through the body as the drive to push ahead [16]. To scientifically show the salp chains, the populace is first isolated to two gatherings: leader and followers. The leader is the salp at the

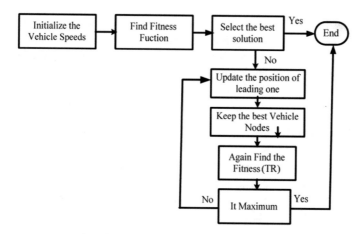

Fig. 2 Flowchart for SA

front of the chain, while the remainders of salps are considered as followers and this flowchart appears in Fig. 2. The Updating state of Salp Algorithm (SA) is

$$O_j^1 = \begin{cases} g_j + W_1[(H_j - L_j)W_2 + L_j] & W_3 \geq 0 \\ g_j - W_1[(H_j - L_j)W_2 + L_j] & W_3 < 0 \end{cases} \tag{5}$$

Here, O_j^1 signifies the position of the first salp (leader) in jth dimension, the position of the food source in jth dimension is symbolized as f_i; the upper bound and lower bound is indicate as H_j and L_j; W_1, W_2, and W_3 symbolizes random number randomly generated in the interval of [0,1].

To update the position of the supporters, Newton's law of movement is used. The inconsistency between iterations is equivalent to 1, and considering the underlying speed as 0, this condition can be communicated as pursues:

$$O_j^i = 0.5[O_j^i + O_j^{i-1}] \tag{6}$$

Salp chain is very liable to find a superior solution by exploring and exploiting the space around it. In this manner, the salp chain can possibly move towards the global optimum that changes through the course of iterations.

Optimal Path Selection: From the above-Updated method the better path will be chosen, to get the most extreme Transmission rate, so the target characterized as

$$\text{Fitness} = \text{MAX}(\text{Transmission Rate}) \tag{7}$$

In light of the optimization model, we are ascertaining the mobility by NMP, optimal route for information dispersal in inadequate just as a dense network of vehicles. For this, we chose to pick a course that minimized the probability of the event of obstructions. VANET routing is the way toward finding the optimal path among source and destination node and afterward sending a message in a coordinated way. The dynamicity in topology mobility models and the execution measurements of the network interest for a dynamic routing protocol.

4 Result Analysis

Evaluation of our proposed NMP-SA model compared to Optimal Link state Protocol (OLSR) and AODV with SA models, its implemented in Network simulator with 4 GB RAM, i7 processor system, Moreover the analysis considered parameters are Packet To Delivery Ratio, Delay, Transmission Rate and Throughput based on Vehicle varying.

Table 1 explains the parameters which we are taken for the analysis. The proposed VANET communication is implemented in Network simulator 3. The chosen routing protocol for the efficient route selection process is NMP, considered number of vehicles is 200, packet size is 150 bytes and IDM is for mobility model analysis.

Table 2 explains the performance analysis of VANET communication by considering the parameters like PDR, throughput, EED, TR for the number of vehicles (50–200). On comparing the three routing protocols OLSR-SA, AODV-SA and proposed NMP-SA, the NMP-SA achieves optimal result in all parameters.

Figure 3a demonstrates the Packet to delivery Ratio (PDR) versus mobility and these parameters are analyzed and compared with three different routing protocols along with optimization such as OLSR-SA, AODV-SA and proposed NMP-SA. On comparing the proposed model with the existing approach, NMP protocol with SA optimization attains high PDR.

Figure 4 shows network throughput of OLSR-SA, AODV-SA and NMP-SA protocols along with number of mobility nodes parkway based situation. Here we may understand that the throughput of OLSR-SA protocol begin at around 1000 Kb/s and with the expansion in number of hubs, the throughput increments bit by bit near its lower esteem, I-e around 2500 Kb/s. Despite the fact that the NMP-SA is considered as an effective routing protocol yet here if there should arise an occurrence of VANET higher node mobility its execution decreases by methods for lower throughput.

Table 1 Simulation parameters	Parameter type	
	Network simulator	3
	Routing protocol	NMP
	No. of vehicles	200
	Packet size	150 bytes
	Mobility model	IDM

Table 2 Comparison of performance measures

No. of vehicle	Parameters	OLSR-SA	AODV-SA	NMP-SA
50	PDR (%)	65.41	69.14	79.41
100		66.85	72.46	76.11
150		73.46	79.22	83.41
200		79.82	80.11	88.22
50	Throughput (kbps)	1214	1278	2004
100		2154	2645	3254
150		2078	2641	3854
200		2345	3241	4698
50	EED (ms)	22.45	28.42	15.52
100		35.11	32.74	28.55
150		49.741	32.76	36.22
200		56.74	42.22	48
50	TR (%)	62.41	76.48	72.14
100		73.44	77.12	76.45
150		82.11	82.1	89.47
200		86.19	93.45	95.45

Fig. 3 PDR versus mobility

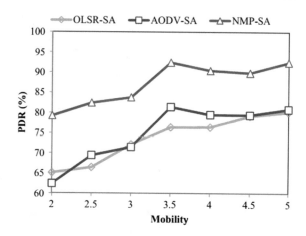

Figure 5 shows the system End-End Delay of OLSR-SA, AODV-SA and NMP-SA protocols along with number of nodes. Here we may understand that the End-End Delay of OLSR-SA convention begin at about 10.2 m s and with the expansion in number of nodes, the End-End Delay increments step by step near its higher esteem, I-e about 40 ms. Despite the fact that the NMP-SA is considered as a decent reactive routing protocol yet here if there should arise an occurrence of VANET higher node mobility its execution decays by methods for higher End-End delay. The essential

Fig. 4 Throughput versus
mobility

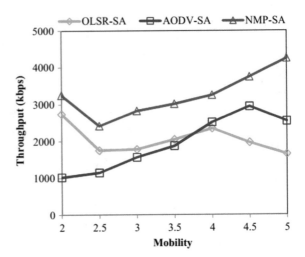

Fig. 5 EED versus mobility

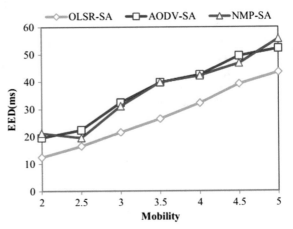

reason of higher End-End Delay is the higher node mobility if there should be an
occurrence of VANETs.

The transmission ratio of different routing protocols like of OLSR-SA, AODV-
SA and NMP-SA are explained in Fig. 6. By varying the mobility nodes 2–5, the
transmission ratio is measured and compared among the routing protocols. The bar
graph concludes that the proposed NMP-SA achieves high TR when compared to
OLSR-SA, AODV-SA.

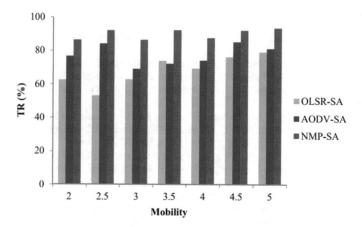

Fig. 6 Transmission ratio versus mobility

References

1. Zhang, L., Lakas, A., El-Sayed, H., Barka, E.: Mobility analysis in vehicular ad hoc network (VANET). J. Netw. Comput. Appl. **36**(3), 1050–1056 (2013)
2. Chahal, M., Harit, S.: Optimal path for data dissemination in vehicular ad hoc networks using meta-heuristic. Comput. Electr. Eng. **76**, 40–55 (2019)
3. Zhu, W., Gao, D., Fong, A.C.M., Tian, F.: An analysis of performance in a hierarchical structured vehicular ad hoc network. Int. J. Distrib. Sens. Netw. **10**(5), 969346 (2014)
4. Jagannath, J., Furman, S., Jagannath, A., Ling, L., Burger, A., Drozd, A.: HELPER: heterogeneous efficient low power radio for enabling ad hoc emergency public safety network (2019). arXiv:1903.08974
5. Fahad, M., Aadil, F., Khan, S., Shah, P.A., Muhammad, K., Lloret, J., Wang, H., Lee, J.W., Mehmood, I.: Grey wolf optimization based clustering algorithm for vehicular ad-hoc networks. Comput. Electr. Eng. **70**, 853–870 (2018)
6. Ding, Z., Ren, P., Du, Q.: Ant colony optimization based delay-sensitive routing protocol in vehicular ad hoc networks. In: International Conference on Internet of Things as a Service, pp. 138 − 148. Springer, Cham (2018)
7. Lakas, A., Fekair, M.E.A., Korichi, A., Lagraa, N.: A multiconstrained QoS-compliant routing scheme for highway-based vehicular networks. Wirel. Commun. Mobile Comput. 2019
8. Gaikwad, D.S. and Zaveri, M.: VANET routing protocols and mobility models: a survey. In: Trends in Network and Communications, pp. 334 − 342. Springer, Berlin (2011)
9. Zeadally, S., Hunt, R., Chen, Y.S., Irwin, A., Hassan, A.: Vehicular ad hoc networks (VANETS): status, results, and challenges. Telecommun. Syst. **50**(4), 217–241 (2012)
10. Radaur, T.J.: Quality of Service and Scalability in Vehicular Ad Hoc Networks
11. Wahid, I., Ikram, A.U.A., Ahmad, M., Ullah, F.: An improved supervisory protocol for automatic selection of routing protocols in environment-aware vehicular ad hoc networks. Int. J. Distrib. Sens. Netw. **14**(11), 1550147718815051 (2018)
12. Mirjalili, S., Gandomi, A.H., Mirjalili, S.Z., Saremi, S., Faris, H., Mirjalili, S.M.: Salp swarm algorithm: a bio-inspired optimizer for engineering design problems. Adv. Eng. Softw. **114**, 163–191 (2017)
13. Chen, Y.S., Hsu, C.S., Cheng, C.H.: Network mobility protocol for vehicular ad hoc networks. Int. J. Commun Syst **27**(11), 3042–3063 (2014)

14. Raw, R.S., Toor, V., Singh, N.: Comprehensive study of estimation of path duration in vehicular ad hoc network. In: Advances in Computing and Information Technology, pp. 309 − 317. Springer, Berlin (2013)

15. Kaur, S., Aseri, T.C. and Rani, S.: QoS-Aware routing in vehicular ad hoc networks using ant colony optimization and bee colony optimization. In: Proceedings of 2nd International Conference on Communication, Computing and Networking, pp. 251 − 260. Springer, Singapore (2019)

16. Halim, A.H.A., Warip, M.M., Ahmad, R.B., Elias, S.J.: Optimization of vehicular ad hoc network using taguchi method. In: 2015 International Conference on Computer, Communications, and Control Technology (I4CT), pp. 147 − 151. IEEE (2015)

17. Umer, T., Amjad, M., Shah, N., Ding, Z.: Modeling vehicles mobility for connectivity analysis in VANET. In: Intelligent Transportation Systems, pp. 221 − 239. Springer, Cham (2016)

18. Shankar, K., Elhoseny, M., Chelvi, E.D., Lakshmanaprabu, S.K., Wu, W.: An efficient optimal key based chaos function for medical image security. IEEE Access **6**, 77145–77154 (2018)

19. Elhoseny, M., Shankar, K., Lakshmanaprabu, S. K., Maseleno, A., Arunkumar, N.: Hybrid optimization with cryptography encryption for medical image security in Internet of Things. In: Neural Computing and Applications, pp. 1 − 15. (2018) https://doi.org/10.1007/s00521-018-3801-x

20. Shankar, K., Elhoseny, M., Kumar, R.S., Lakshmanaprabu, S.K., Yuan, X.: Secret image sharing scheme with encrypted shadow images using optimal homomorphic encryption technique. J. Ambient Intell. Humanized Comput. 1 − 13 (2018). https://doi.org/10.1007/s12652-018-1161-0

21. Gaber, T., Abdelwahab, S., Elhoseny, M., Hassanien, A.E.: Trust-based secure clustering in WSN-based intelligent transportation systems. Comput. Netw. https://doi.org/10.1016/j.comnet.2018.09.015 (2018). Accessed 17 Sept 2018

22. Mohamed, R.E., Ghanem, W.R., Khalil, A.T., Elhoseny, M., Sajjad, M., Mohamed, M.A.: Energy efficient collaborative proactive routing protocol for wireless sensor network. Comput. Netw. https://doi.org/10.1016/j.comnet.2018.06.010 (2018). Accessed 19 June 2018

23. Elhoseny, Mohamed, Tharwat, Alaa, Yuan, Xiaohui, Hassanien, A.E.: Optimizing K-coverage of mobile WSNs. Expert Syst. Appl. **92**, 142–153 (2018)

24. Elsayed, Walaa, Elhoseny, Mohamed, Sabbeh, Sahar, Riad, Alaa: Self-maintenance model for wireless sensor networks. Comput. Electr. Eng. **70**, 799–812 (2018)

25. Elhoseny, M., Tharwat, A., Farouk, A., Hassanien, A.E.: K-coverage model based on genetic algorithm to extend WSN lifetime. IEEE Sens. Lett. **1**(4), 1 − 4 (2017). IEEE

26. Elhoseny, M., Farouk, A., Zhou, N., Wang, M.-M., Abdalla, S., Batle, J.: Dynamic multi-hop clustering in a wireless sensor network: performance improvement. Wirel. Pers. Commun. **95**(4), 3733 − 3753

27. Elhoseny, M., Yuan, X., Yu, Z., Mao, C., El-Minir, H., Riad, A.: Balancing energy consumption in heterogeneous wireless sensor networks using genetic algorithm. IEEE Commun. Lett. IEEE **19**(12), 2194–2197 (2015)

Studying Connectivity Probability and Connection Duration in Freeway VANETs

Sherif M. Abuelenin and Adel Y. Abul-Magd

Abstract This chapter discusses the connectivity in freeway vehicular ad hoc networks (VANETs). Estimating connectivity is necessary to measure the effectiveness of vehicular communication. The connection duration between any two vehicles is a function of their relative velocity, and the connectivity probability is a function of the headway distribution. Both are functions of the traffic regime. We begin by presenting some recent research results in estimating the connectivity probability and the communication duration between vehicles in different traffic conditions, namely single-lane same direction traffic. We then extend the results to other scenarios.

Keywords Intervehicle communications · VANETs · Connectivity probability

1 Introduction

Vehicular communications play an essential role in intelligent transportation systems [1], with applications varying from safety and driver assistant systems to providing infotainment and internet access [2–4]. In vehicular communications, vehicles are equipped with radio devices to enable information exchange among them [5–8]. Standards for vehicular communications are emerging [9, 10], this includes IEEE 802.11p (a.k.a. WAVE) [11–13], IEEE 1609, OSI CALM-M5, and ETSI ITS-G5. Vehicular ad hoc networks (VANETs) are considered a special type of mobile ad hoc networks (MANETs). In VANETs, information is exchanged spontaneously (in ad hoc manner), without the need of infrastructure. Road-side units can also be used to improve communications in VANETs [14–16]. With increasing numbers of vehicles being connected to the Internet of Things (IoT), the conventional VANETs

S. M. Abuelenin (✉)
Department of Electrical Engineering, Faculty of Engineering, Port-Said University, Port-Fouad, Port-Said 42526, Egypt
e-mail: s.abuelenin@gmail.com

A. Y. Abul-Magd (Deceased)
Department of Basic Mathematics, Faculty of Science, Zagazig University, Zagazig 44519, Egypt

© Springer Nature Switzerland AG 2020
M. Elhoseny and A. E. Hassanien (eds.), *Emerging Technologies for Connected Internet of Vehicles and Intelligent Transportation System Networks*, Studies in Systems, Decision and Control 242, https://doi.org/10.1007/978-3-030-22773-9_3

are developing into the Internet of Vehicle (IoV) [17, 18]. IoV is an integrated network system in which VANETs are an essential component [18].

Unlike MANETs in general, the nodes in VANETs (i.e. vehicles) are highly mobile [8] resulting in a rather fast change in network topology. The movement of the nodes is constrained to well defined areas (i.e. roads), they have restricted directions of motions, and the spacings between them follow certain spatial distributions that are well studied and modelled within the framework of traffic theory. These characteristics result in high probability of network partitioning and no guarantee of end-to-end connectivity [3], however, they also ease studying certain characteristics of VANETs such as connectivity. Studying connectivity features is fundamental aspect in planning vehicular networks [10, 16].

Communication characteristics in VANETs are directly affected by the mobility and spatial distribution of vehicles. This chapter considers two aspects; the connectivity probability and the connection duration in freeway VANETs. The chapter builds on the authors' previous works [19–21]. In Sect. 2 the problem of estimating the connectivity probability in highway VANETs is presented, along with some special cases. Section 3 discusses the estimation of the connection duration between vehicles. The chapter is concluded in Sect. 4.

2 Instantaneous Connectivity of VANETs

The connectivity probability in a VANET can be defined as the probability that a (single-hop or multi-hop) communication path exists between the first and the last vehicle in the considered road segment. Because of the continuous movement of nodes which consistently affect the connectivity, we refer to the connectivity at a given moment as the instantaneous connectivity. A very common method of estimating the connectivity is to use the disc model [5–10, 19–22]. The model assumes that the communication range, denoted R, of all vehicles within the network is fixed. If the distance between any two vehicles (the headway) is less than R, the two vehicles are considered connected. The model is widely used, and it captures the average behaviour of the network [10]. In this model, the communication range R defined as the distance that guarantees a certain level of received power (or alternatively, a certain level of signal-to-noise ratio) at the receiver. Using Friis equation, $R = K \cdot \sqrt[\alpha]{\frac{P_t}{P_r}}$, where α is the pathloss exponent.

To estimate the connectivity probability of a VANET using the disc model, we let the distance between the two vehicles V_i and V_{i+1} be denoted by the random variable X_i, $i = 1, 2, \ldots, N-1$ [5, 6], where N is the number of vehicles in the considered road segment. A VANET is connected if and only if X_i is less than R, for all values of i. Hence, the connectivity probability depends on the distribution of the intervehicle distances. The probability of connectivity P_c is given by [5–7];

$$P_c = \prod_{i=1}^{N-1} P(X_i < R) = \prod_{i=1}^{N-1} F_X(R) = F_X(R)^{N-1} \tag{1}$$

where FX(x) is the headway cumulative distribution function (CDF).

It is common in traffic studies to assume that the distances between vehicles are exponentially distributed;

$$F_X(x) = \left(1 - e^{-\rho x}\right) u(x) \tag{2}$$

where $u(x)$ is the Heaviside unit step function, and ρ is the average traffic density in vehicles per unit distance (and is equal to the reciprocal of the mean headway distance).

However, this assumption is not accurate for all traffic scenarios. According to Kerner's three phase traffic theory [23], traffic can be either in free-flow phase (light traffic), synchronized flow (in-between), and wide moving jam (highly congested). In free-flow traffic, drivers are free to choose their own speeds [24], and cars can move independent of each other [25]. Accordingly, the distance between any two cars is uncorrelated, and X_i's are independent and identically distributed [5, 6]. Therefore, the exponential headway distribution, as described in Eq. (2), is valid in (multi-lane) free-flow regimes, where the lane separation between vehicles is much less than the headway in free-flowing traffic. Ignoring the lane separation between vehicles allows for the headway distance to be zero (e.g. for two adjacent vehicles), as the exponential distribution suggests. Recent studies confirmed that this assumption is indeed valid in light traffic (free flow phase) [5, 26]. As for other traffic regimes and scenarios, several other distributions were proposed to provide better modelling of empirical data of headway distribution. A survey of headway models can be found in [26]. The alternative headway distribution models include the generalized-extreme-value (GEV) and lognormal distributions to model headway distribution in synchronized flow phase [5, 26]. Theoretical models were also presented from different physical perspectives, e.g. Gaussian unitary ensemble (GUE) and super-statistics were used to model headway distribution in traffic jam and in the synchronized-flow phase [24].

Connectivity analysis based on the GEV model [5] confirmed that the connectivity probability estimation is sensitive to the traffic headway model. In the rest of this section we consider two interesting cases; free-flow traffic in single-lane highways [20, 21], and wide (moving) traffic jam.

2.1 Effect of Safe Driving Distance in Free-Flow Traffic

Under light traffic conditions (free flow) drivers can choose their own speed. The distances between cars are uncorrelated. Intervehicle distances are traditionally modeled using exponential distribution, as described by Eq. (1). This distribution model permits zero distance between successive vehicles, which is impossible. Vehicles must keep a minimum—safe—distance from each other. We proposed, [20], to use the slightly modified shifted-exponential distribution model;

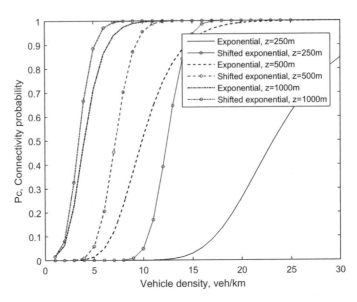

Fig. 1 Probability of connectivity using both exponential and shifted exponential distribution models, each simulated using three communication ranges; z = 250, 500, and 1000 m, the road segment length $L = 10$ km

$$F(x) = \left(1 - e^{-\tilde{\rho}(x-b)}\right)u(x - b) \qquad (3)$$

In order to avoid the zero-headway distance problem allowed in the exponential model, the shift parameter b in Eq. (2) is used to model the summation of the minimum inter-vehicle distance and the average vehicle length. Use of this model was justified in [20] by applying both models to empirical real traffic data. Root mean square error (RMSE) analysis verified that this model provides a better fit for single-lane light traffic headway data.

It is noted that, in Eq. (3), $\tilde{\rho}$ does not represent the traffic density. The mean value of the shifted-exponential distribution whose CDF is given by (3) equals $b + 1/\tilde{\rho}$. Therefore, the relation between the parameter $\tilde{\rho}$ and the actual traffic density is given by $\tilde{\rho} = \rho/(1 - b\rho)$. Applying the modified distribution to Eq. (1) we can find the connectivity probability.

Figure 1 shows the connectivity probability using both models three different values of the communication range. The parameters used in obtaining these results are as follows; the road segment $L = 10$ km, and $b = 39.1$ m, as obtained from data fitting [20]. The results interestingly show that the minimum headway distance will result in increasing the probability of connectivity in single-lane free-flow VANETs. It is noted here that these analysis, in consistence with our assumptions, are valid only for light traffic scenarios, i.e. $\rho < 20$ vehicles per km [27, 28], and therefore large communication ranges must be utilized.

2.2 Connectivity in Wide Traffic Jam Conditions

Another interesting case to consider is wide (moving) traffic jam. The independence of safe distance on the traffic density may not be the case under congested traffic regimes, where higher density will force the drivers to keep smaller distances between the vehicles. In Kerner's [23] traffic theory, congested traffic is classified into two distinct phases: synchronized flow and wide moving jams. In synchronized flow, the speeds of the vehicles are low and vary quite a lot between vehicles, but the traffic flow remains close to free flow [24]. In wide moving jams, vehicle speeds are more equal and lower. The term 'wide (moving) jams' is used as distinction from extreme traffic congestion (traffic jam), where vehicles are fully stopped for periods of time. It was shown [24, 29] that intervehicle distances in wide moving jams can be well modeled using the distribution of the nearest-neighbor spacing distribution of Gaussian unitary ensembles (GUE) of random matrix theory (RMT). The probability density function (PDF) of the distribution is described by Eq. (4).

$$p(x) = \frac{32}{\pi^2} \frac{x}{D^3} e^{-\frac{4x^2}{\pi D^2}} \cdot u(x) \tag{4}$$

where D is the mean distance between vehicles (reciprocal of the traffic density). Unlike the exponential distribution, the tail of this distribution decays exponentially with the square of the random variable x. This means that the probability of finding a distance that is much larger than the mean distance is highly unlikely. This agrees with the nature of wide traffic jams. Reference [22] analyzed both distributions and showed that—indeed as expected—the node degree distribution for GUE distances shows a much narrower node degree distribution in comparison to that of the exponential distribution.

In wide jams, one expects that a much smaller communication range would be needed to guarantee connectivity between vehicles. Figure 2, below, shows the probability of connectivity using the GUE model in comparison to using the exponential model. The curves in the figure are calculated using a communication range of 25 m. As one would expect, the connectivity probability estimated using the GUE model is much higher than that estimated using the exponential headway distribution model.

In this section we studied the instantaneous probability of connectivity in VANETs and showed how the selection of the appropriate headway distribution model is necessary for estimating this probability. Using different models can greatly affect the estimation results.

3 Connection Duration Estimation

The probability distribution of the duration of the connection has also been a subject of research [19, 22]. In VANETs, connection duration between any two vehicles is a function of the communication range and the relative velocity between them.

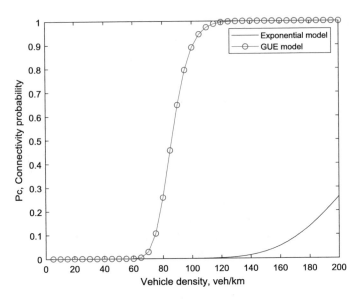

Fig. 2 Probability of connectivity in wide jam using both the exponential and GUE distribution models, the communication range is set to be 25 m, and the road segment length $L = 1$ km

Therefore, the probability distribution of this duration is a function of the distribution of the relative velocity between vehicles, which depends on the velocity distribution of each vehicle. It is widely accepted that the velocity of moving vehicles is very well modelled by the Gaussian distribution. This has been supported by theoretical modelling as well as empirical data (e.g. [30, 31]). In this section, we show that [19] in some cases, vehicles velocity cab be better modelled by other distributions, and that this affects the estimation of the connection duration.

Figure 3 shows the mean velocity in every hour of the one day as computed using real traffic data. The data is collected from dual-loop sensors on the interstate I-80 road in California and are obtained from Berkeley Highway Laboratory (BHL) project [32]. The data are collected using dual loop sensor stations installed on the five lane Different time periods are selected to represent different traffic regimes, with each time period lasting for one hour. We empirically find the CDF of each. Fitting the CDF to common distributions showed that in most cases, the velocity is indeed Gaussian distributed, except for the two hours in which there was a transition between high and low average speeds, i.e. hours 15 and 18. In these two times of the day, the GEV and the lognormal distribution showed better fir with the empirical data, as shown in Fig. 4.

To study the connection duration probability distribution, we need first to determine the distribution of the relative velocity between two vehicles. If the velocity of each vehicle is considered as a random variable (RV), we need to find the distribution of the difference between the two RVs. Reference [22] has studied the connection duration distribution using the assumption of Gaussian distributed velocities. Here,

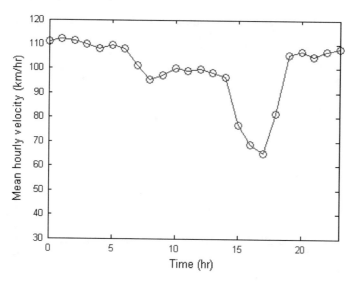

Fig. 3 Mean hourly velocity in 24 h period

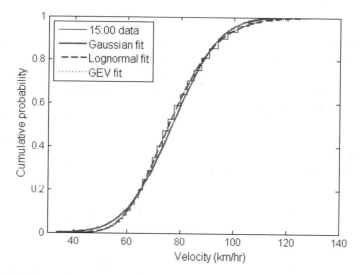

Fig. 4 CDF plot of traffic velocities at 3:00 pm

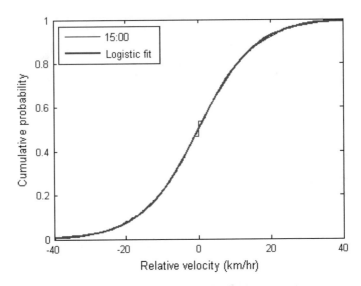

Fig. 5 Plots of relative vehicles velocities distribution; empirical and logistic

we show the effect of the different velocity distribution (of the identified transition times) on the connection duration. Finding the distribution of the difference between two lognormal RVs is a challenging problem [33]. Therefore, we limit the analysis of this section to modelling the considered velocity using GEV.

The GEV distribution has three parameters; μ for location, σ for scale, and k for shape. The described fitting results, for 3 pm data, show that k is close to zero, which means that the distribution is almost a Gumbel (GEV-type-I) distribution. The difference between two Gumbel distributed RVs follows a logistic distribution [34]. The logistic distribution has the following PDF;

$$p(v) = \frac{e^{-\frac{v-\mu}{\sigma}}}{\sigma\left(1 + e^{-\frac{v-\mu}{\sigma}}\right)^2} = \frac{1}{4\sigma}\operatorname{sech}^2\left(\frac{v-\mu}{2\sigma}\right) \tag{5}$$

Figure 5 shows that the CDF of the logistic distribution matches the empirical CDF of the relative velocity data at 3 pm.

The connection duration between two vehicles in a VANET is $t_c = 2R/|\Delta v|$, where R is the communication range and Δv is the relative velocity. This equation has two solutions; $\Delta v = \pm 2R/t_c$.

The random variable t_c is a function of the random variable Δv. The distribution of t_c can be easily found using the distribution of Δv [35]. If the distribution PDF of the velocity difference is known to be $p_v(v)$, (and assuming that the lane separation between vehicles is much less than the headway) then, the distribution PDF of connectivity duration 't_c' can be found to be [22];

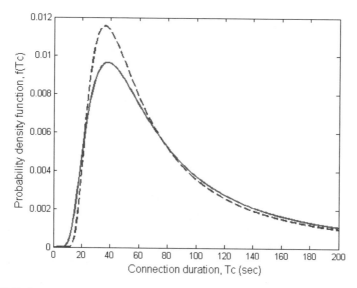

Fig. 6 PDF of connection duration using Gaussian distribution (blue, dashed) and logistic distribution (red, solid)

$$p_{T_C}(t_c) = \frac{2R}{t_c^2}\left(p_v\left(\frac{2R}{t_c}\right) + p_v\left(-\frac{2R}{t_c}\right)\right)u(t_c) \tag{6}$$

Substituting (5) in (6) gives an estimation of the connection duration distribution for the GEV velocity, as plotted in Fig. 6 in comparison to that obtained using Gaussian distribution.

The figure shows that, for the same transmission range ($R = 100$ m), the probability of having two vehicles communicating for certain duration is higher when estimated using Logistic distribution.

The results provided so far in this section considers the connection duration of two vehicles going along the same direction of the road. It would be interesting to consider extending the results to the case when both traffic directions are utilized in a bi-directional VANET. In this case, the relative velocity between vehicles of opposing directions greatly increases. The relative velocity now is defined as the summation, not the difference, of the two cars velocities. Such scenario has been studied for Gaussian velocity distribution [22]. However, to extend it using the two alternative distributions considered above (Gumbel and lognormal) case is a very complicated problem. Unlike obtaining the distribution of the difference between two Gumbel RVs, finding the distribution of the summation of two Gumbel RVs turns out to be challenging [36], with no simple mathematical formula that can represent the difference. The same is true with finding the sum of two lognormal RVs (which, as we showed above, can be used to model the same problem). However, several approximations were provided for such a problem. E.g. Ref. [37] deduced a "highly accurate simple closed-form approximations to lognormal sum". Such approximations may be used to study the connection duration in the very specific case described here. However, this requires obtaining empirical data for both sides of the roads at the same time for verification. This problem is left for future consideration.

4 Conclusions

In this chapter we discussed how the communication aspects, namely the instantaneous connectivity probability and the connection duration, of freeway VANETs are affected by the traffic regime and its characteristics. Estimating connectivity is necessary to measure the effectiveness of vehicular communication. We used two examples, the single-lane free-flow traffic and the wide moving jam, to show that the estimated probability of establishing a connection between vehicles in varies greatly according to the headway distribution model. Care must be taken when choosing the appropriate distribution at any given scenario to avoid underestimation (or overestimation) of connectivity probability. We Also discussed how the probability distribution of the connection duration is affected by the velocity distribution of vehicles.

References

1. Lu, N., Cheng, N., Zhang, N., Shen, X., Mark, J.W.: Connected vehicles: solutions and challenges. IEEE Internet Things J. 1(4), 289–299 (2014). https://doi.org/10.1109/JIOT.2014. 2327587
2. Campolo, C., Molinaro, A., Vinel, A., Zhang, Y.: Modeling and enhancing infotainment service access in vehicular networks with dual-radio devices. Veh. Commun. 6, 7–16 (2016)
3. Toor, Y., Muhlethaler, P., Laouiti, A., La Fortelle, A.D.: Vehicle ad hoc networks: applications and related technical issues. IEEE Commun. Surv. Tutor. 10(3), 74–88 (2008). Third Quarter
4. Shrestha, R., Bajracharya, R., Nam, S.Y.: Challenges of future VANET and cloud-based approaches. Wireless Commun. Mob. Comput. 2018, Article ID 5603518. https://doi.org/10. 1155/2018/5603518
5. Cheng, L., Panichpapiboon, S.: Effects of intervehicle spacing distributions on connectivity of VANET: a case study from measured highway traffic. IEEE Commun. Mag. 50(10), 90–97 (2012)
6. Panichpapiboon, S., Pattara-Atikom, W.: Connectivity requirements for self-organizing traffic information systems. IEEE Trans. Veh. Technol. 57(6), 3333–3340 (2008)
7. Panichpapiboon, S., Pattara-Atikom, W.: Connectivity requirements for a self-organizing vehicular network. In: 2008 IEEE Intelligent Vehicles Symposium, Eindhoven, pp. 968–972 (2008). https://doi.org/10.1109/ivs.2008.4621219
8. Busson, A.: Analysis and simulation of a message dissemination algorithm for VANET. Int. J. Commun. Syst. (2011)
9. Gramaglia, M., Trullols-Cruces, O., Naboulsi, D., Fiore, M., Calderon, M.: Mobility and connectivity in highway vehicular networks: a case study in Madrid. Comput. Commun. 78, 28–44 (2015)
10. Naboulsi, D., Fiore, M.: Characterizing the instantaneous connectivity of large-scale urban vehicular networks. IEEE Trans. Mob. Comput. 16(5), 1272–1286 (2017). https://doi.org/10. 1109/TMC.2016.2591527

11. IEEE P802.11p/D6.01, Part 11: wireless LAN medium access control (MAC) and physical layer (phy) specifications-amendment 7: wireless access in vehicular environments (2009)
12. IEEE Std, IEEE standard for wireless access in vehicular environments (WAVE) multi-channel operation (2010)
13. IEEE family of standards for wireless access in vehicular environments (WAVE)—IEEE 1609 series. IEEE (2013)
14. Sou, S.I., Tonguz, O.K.: Enhancing VANET connectivity through roadside units on high-ways. IEEE Trans. Veh. Technol. 60(8), 3586–3602 (2011). https://doi.org/10.1109/TVT.2011. 2165739
15. Aslam, B., Amjad, F., Zou, C.: Optimal roadside units placement in urban areas for vehicular networks. In: Proceedings of the IEEE Symposium on Computers and Communications, pp. 423–429 (2012)
16. Durrani, S., Zhou, X., Chandra, A.: Effect of vehicle mobility on connectivity of vehicular ad hoc networks. In: 2010 IEEE 72nd Vehicular Technology Conference - Fall, Ottawa, ON, pp. 1–5 (2010). https://doi.org/10.1109/vetecf.2010.5594505
17. Kaiwartya, O., Abdullah, A.H., Cao, Y., Altameem, A., Prasad, M., Lin, C.T., Liu, X.: Internet of vehicles: motivation, layered architecture, network model, challenges, and future aspects. IEEE Access 4, 5356–5373 (2016)
18. Yang, F., Wang, S., Li, J., Liu, Z., Sun, Q.: An overview of internet of vehicles. China Commun. 11(10), 1–15 (2014)
19. Abuelenin, S.M., Abul-Magd, A.Y.: Empirical study of traffic velocity distribution and its effect on VANETs connectivity. In: 2014 International Conference on Connected Vehicles and Expo (ICCVE), Vienna, pp. 391–395 (2014). https://doi.org/10.1109/iccve.2014.7297577
20. Abuelenin, S.M., Abul-Magd, A.Y.: Effect of minimum headway distance on connectivity of VANETs. AEU-Int. J. Electron. Commun. 69(5), 867–871 (2015)
21. Abuelenin, S.M., Abul-Magd, A.Y.: Corrigendum to effect of minimum headway distance on connectivity of VANETs. AEU-Int. J. Electron. Commun. 83, 566 (2018)
22. Nagel, R.: The effect of vehicular distance distributions and mobility on VANET communications. In: 2010 IEEE Intelligent Vehicles Symposium, San Diego, CA, pp. 1190–1194 (2010). https://doi.org/10.1109/ivs.2010.5547971
23. Kerner, B.S.: The Physics of Traffic: Empirical Freeway Pattern Features, Engineering Applications, and Theory. Springer, Berlin (2004)
24. Abul-Magd, A.Y.: Modeling highway-traffic headway distributions using superstatistics. Phys. Rev. E 76(5), 057101 (2007)
25. Luttinen, R.: Statistical Analysis of Vehicle Time Headways. Teknillinen korkeakoulu, Otaniemi (1996)
26. Li, L., Chen, X.M.: Vehicle headway modeling and its inferences in macroscopic/microscopic traffic flow theory: a survey. Transp. Res. Part C: Emerg. Technol. 76, 170–188 (2017)
27. Abuelenin, S.M., Abul-Magd, A.Y.: Moment analysis of highway-traffic clearance distribution. IEEE Trans. Intell. Transp. Syst. 16(5), 2543–2550 (2015). https://doi.org/10.1109/TITS.2015. 2412117
28. Ayres, T.J., Li, L., Schleuning, D., Young, D.: Preferred time-headway of highway drivers. In: ITSC 2001. 2001 IEEE Intelligent Transportation Systems. Proceedings (Cat. No.01TH8585), Oakland, CA, pp. 826–829 (2001)
29. Krbalek, M., Seba, P.: The statistical properties of the city transport in cuernavaca (Mexico) and random matrix ensembles. J. Phys. A 33(L229–L234) (2000)
30. Krbálek, M., Kittanová, K.: Lattice thermodynamic model for vehicular congestions. Procedia-Soc. Behav. Sci. 20, 398–405 (2011)
31. Krbálek, M.: Theoretical predictions for vehicular headways and their clusters. J. Phys. A: Math. Theor. 46(44), 445101 (2013)
32. http://bhl.path.berkeley.edu/
33. Lo, C.F.: The sum and difference of two lognormal random variables. J. Appl. Math. (2012)
34. Gumbel, E.J.: Statistics of Extremes. Courier Dover Publications (2012)

35. Papoulis, A., Pillai, S.U.: Probability, Random Variables, and Stochastic Processes. Tata McGraw-Hill Education (2002)
36. Nadarajah, S.: Linear combination of Gumbel random variables. Stoch. Env. Res. Risk Assess. **21**(3), 283–286 (2007)
37. Beaulieu, N.C., Rajwani, F.: Highly accurate simple closed-form approximations to lognormal sum distributions and densities. IEEE Commun. Lett. **8**(12), 709–711 (2004)

A Survey of Different Storage Methods for NGN Mobile Networks: Storage Capacity, Security and Response Time

Boughanja Manale and Mazri Tomader

Abstract Following the considerable evolution of new technologies, as well as the computerization of different exchanges and gadgets in lot of areas lately, and with the appearance of the new generation 5G, the quantity of connected devices will increase in expanded way. Similarly, the creation of data in its different structures will be important. Consequently, the value of studying the distinctive storage methods. This paper introduces the different stockpiling techniques, which enables average readers to have an overview of the distinctive methods including the definition and architecture.

Keywords 5G · Storage method · Security · Next generation network (NGN)

1 Introduction

These days cell phones (e.g. advanced mobile phone, tablet and PC) are getting to be a fundamental piece of human life, as an instrument of communication. Mobile users are continuously creating data from different versatile applications (e.g. IOS, Android applications [1–6]). Hence the emergence of the new era of Big Data. The term big data is defined as a set of voluminous and complex data, characterized by five measurements: volume, that represent the data created, velocity, or speed depicts the recurrence with which data is produced, caught and shared; variety, indicate the heterogeneity of data sources; veracity, which mean the reliability and quality of data and value, show the data inference. With the expansion of the age of information the limits of conventional stockpiling have begun less intriguing, so scientists are endeavoring to utilize other techniques that resolve this issue. Yet, many researches

B. Manale (✉) · M. Tomader
Electrical Systems and Telecommunications Engineering
Ibn Tofail Science University, Kenitra, Morocco
e-mail: boughnja.manale@gmail.com

M. Tomader
e-mail: tomader20@gmail.com

© Springer Nature Switzerland AG 2020
M. Elhoseny and A. E. Hassanien (eds.), *Emerging Technologies for Connected Internet of Vehicles and Intelligent Transportation System Networks*, Studies in Systems, Decision and Control 242, https://doi.org/10.1007/978-3-030-22773-9_4

focus on data storage. The objective to this paper is to look for the different storage methods to deal with the data generated. The strength of a method lies in the capacity of storage, without forgetting also that the security side must be integrated so as not to have losses in our data. Therefore, we have focused on the three most important criteria for versatility of a stockpiling method, namely security, response time and storage capacity. This paper presents a comprehensive survey on distinctive stockpiling systems; to this point we present an investigation of those strategies.

For that our paper is organized as follows: in the Sect. 1 we will present an overview of the state of art of art existing of solutions of the mobile network evolution, Sect. 2 will introduce the distinctive storage methods for mobile network, Sect. 3 we will present existing works. In the last section, we will examine the outcomes of this research the analysis of the study.

2   Overview

Till now 4 generation of cellular communication has been embraced [7]. So, what do we mean by mobile telephony? The mobile telephony is a telecommunication infrastructure that makes it conceivable to communicate by mobile device without being connected by cable to a central office. In this section we will explain the different generation of mobile telephony. Beginning with the first generation which depends on analogic FM cellular system since 1980. They have many implementations for the first generation such as Nordic mobile telephone (NMT), American mobile phone system (AMPS) and Total access communication system (TACS). It relies on the frequency division multiple access (FDMA) this method permits the division of frequencies by dispensing a part of frequency range to every client [8]. They have a several disadvantages such as bad voice quality also the equipment was too heavy for the use, the first generation is considered as an incompatible standard from one region to another. Hence the second generation appeared which utilize the digital modulation which can be divided into two-time division multiple access (TDMA), this method uses a time division of the bandwidth, the principle of which is to divide the available time between the different users [9]. Or code division multiple access (CDMA) [7] which depending on the type of multiplexing used. The second generation came with a lot of advantages such as confidentiality of communication and it allow us at the first-time sending messages. With the use of digital coding it improves the voice clarity and lessens noise in the line. The 3G in 2001 which intend to offer rapid information by utilizing an innovation, for example, wideband code division multiple access (W-CDMA) which is a variant of the CDMA technique [7], and fast Packet Access (HSPA) [7]. These technologies allow speeds much faster than those of the previous generation, and allow multimedia uses such as video transmission, mobile TV, video telephony or broadband internet access. Then the 4G is the fourth generation of mobile telephony. It is marked by the arrival of the new LTE (Long Term Evolution) technology, which is characterized by a theoretical throughput of 150 Mbit/s. The principle improvements in 4G are the increase in theoretical maximum

Table 1 Requirement and realities of 1G through 4G cellular systems

Generation	Requirement	Comments
1G	Analog technology	Deployed on 1980s
2G	Digital technology	Deployed on 1990s, the principal Implementation is: GSM, GPRS and EDGE Allow text message service
3G	Technology:CDMA, UMTS	From 2000 to 2010 Allow video call
4G	LTE technology	Capacity rate 100 Mbs–1 Gbs

velocities (better real throughput and better network capacity to deal with traffic) and lower latency times (decreased reaction time for better interactivity). The expected result for the 4G technology is basically the high-quality audio/video streaming over end to end Internet Protocol. The development is condensed on Table 1:

With the expansion in the quantity of users and the huge interest for data generation for cell phones and gadgets, the challenge in the design of the terminals is associated with the administration of exchange between the adaptability of how to utilize the spectrum and required space and power to given platform. New techniques offer structure measurements that enable the system to adjust to the chances and prerequisites of the terminals in a way that will amplify the spectral efficiency and furthermore boost the battery control. Because of developing dimension of acknowledgment of the remote advancements in various fields, challenges and types of wireless systems associated with them are changing [7]. With the headway of 5G a couple of advancements appear with it: millimeter Waves, small cell, massive MIMO, beamforming and full duplex [10]. The following section we will clarify the distinctive innovations underneath.

2.1 The Millimeter Waves

The cell phones and the various electronic gadgets utilize a specific frequency on the radio frequency spectrum, particularly the frequencies underneath 6 GHz, however these frequencies are ending up dynamically full [11]. The limit of the past generation has achieved its utmost, a fifth generation is created to take care of the issue of the cutoff of the frequency spectrum [12]. It can't in this way contain as much data on a similar radio frequency spectrum which cause a break in the association. Along these lines, the arrangement was to utilize the millimetric wave to almost certainly contain an ever-increasing number of electronic devices [11]. The millimeter waves refer to the waves that are somewhere in the range of 30 and 300 GHz, the thought of millimetric is just with respect to the wave-length, which range between of 1 and 10 mm [7].

2.2 Small Cell

A small cell is an umbrella term for administrator controlled, low-powered radio interchanges equipment [11]. Along these lines, we can characterize a small cell as a little power base station which would be extremely near the customary towers and structure a kind of hand-off group to transmit the signs around the obstructions (for example downpour, trees, etc.), to guarantee that mm-wave frequencies will beat any signal attenuation [7].

2.3 Massive MIMO

The term massive MIMO (multiple input multiple output) refer to the utilization of an enormous measure of antenna [12], the quantity of antennas assumes the number of terminals is served. With the utilization of massive MIMO, various clients are served simultaneity. The main issue that is caused with this technology that the data is communicates in all direction at the same time which causing interference.

2.4 Full Duplex

Full duplex communication—a full-duplex radio can transmit and get distinctive packets at the same time. Its interference cancellation hardware and baseband signal processors can adequately isolate the interference from transmitted signs to get ones [7]. The possible benefits of full duplex wireless have as of late driven specialists to investigate how one may construct such a gadget [10]. The fundamental test is decreasing self-interference which is brought about by the massive MIMO the arrangement was to implement another innovation which is the full duplex [13].

2.5 Beamforming

In light of the quantity of antennas that are utilized in 5G the signal is transmitted in an undirected method for the base station to mobile equipment [7], consequently the beamforming technique is require to streamline the conveyance of the needed radio signal associated to cell phones [11]. With the use of conventional antennas, the signal is transmitted to all equipment in the same area as shown in Figs. 1 and 2 after its implementation and demonstrates the usefulness of the utilization of beamforming strategy. So, the beamforming technique is considered as a signaling system for signal orientation.

Fig. 1 Conventional
antenna

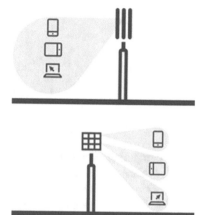

Fig. 2 Beamforming
antenna

3   Storage Methods for Mobile Networks

With the appearance of 5G and its different pillars, the producing information is incrementing particularly. This new technology brings several advantages among them: The volume of connected devices; Energetic efficiency and Reducing response time (no delay). In the previous section we discuss about the new generation and its utility to maintain a strong connectivity and its capability to support a huge quantity of connected devices and applications. For this reason, the need to study the different storage methods is turning into a commitment to manage the information created. In this article; we will investigate the diverse stockpiling techniques for mobile network. The life cycle of a data is as follow: collection, treatment and storage in our case we are focused on the last steep which is the storage. The cloud computing is the first method which appear in our mind when we want to speak about storage. The cloud computing refers to both the applications conveyed as service over the Internet and the equipment and system software in the data center that give those services [14]. In this section we will explain the cloud computing and the other storage method for mobile network.

3.1   Cloud Computing

With the fast advancement of processing and capacity advances and the accomplishment of the Internet, computing resources have turned out to be less expensive, more dominant and more universally accessible than any other time. This technological pattern has empowered the acknowledgment of another figuring model called computing model, in which assets (e.g., CPU and storage capacity) are given as general utilities that can be rented also, discharged by users through the Internet in an on-request fashion. The NIST (National Institute of Standards and Technology) definition of

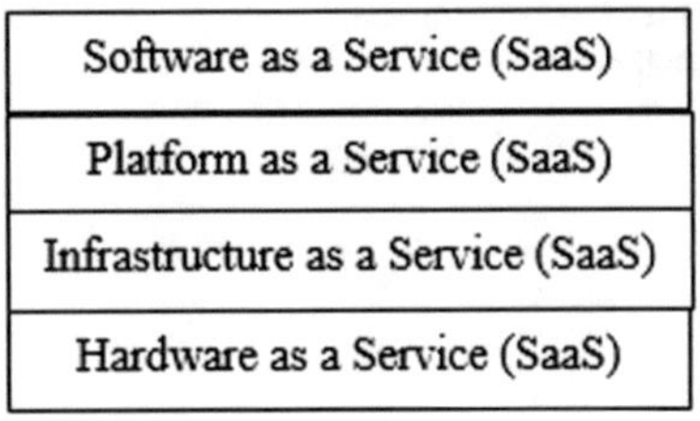

Fig. 3 Cloud layered
architecture [21]

Software as a Service (SaaS)
Platform as a Service (SaaS)
Infrastructure as a Service (SaaS)
Hardware as a Service (SaaS)

cloud computing Cloud computing *is a model for enabling convenient, on-demand network access to a shared pool of configurable computing resources (e.g., net- works, servers, storage, applications, and services) that can be rapidly provisioned and released with minimal management effort or service provider interaction.* The architecture of the cloud is shown in Fig. 3:

The cloud services are generally classified on a layer idea (Fig. 3). In the upper layers of this paradigm, Infrastructure as a Service (IaaS), Platform as a Service (PaaS), and Software as a Service (SaaS) are stacked.

- IaaS: gives the required infrastructure as a service. The customer need not purchase the required servers, data center or the network resources.
- PaaS: give a computing platform utilizing the cloud infrastructure. It has all the application regularly required by the customer deployed on it. Hence the customer need not experience the problems of purchasing and installing the software and hardware required for it [15]. For examples are GAE, Microsoft's Azure [16].
- SaaS: over the internet, along these lines dispensing with the need to introduce and run the application on the users system [17]. Essential qualities of this are: [18] Network-based access and the board of economically accessible programming that are overseen from incorporated areas and empowering clients to get to these applications remotely through the internet.

3.2   Mobile Cloud Computing

In [19] describe MCC as another worldview for versatile applications whereby the data processing and capacity are moved from the cell phone to ground-breaking and centralized computing platforms situated in clouds. These centralized applications are then gotten to over the remote association dependent on a thin native client or internet browser on the cell phones. Along these lines, we can say that the MCC give both the data processing and the storage capacity in the cloud. The essential design of MCC is progressively similar to the cloud computing, the cloud gives a lot of storage capacity and administration while the remote customers get administration as indicated by their need as is demonstrated is Fig. 4.

We can separate the architecture of MCC in two types: agent-client scheme and collaborated scheme [20]. In the next paragraph we will explain the two types.

Fig. 4 Basic architecture of MCC

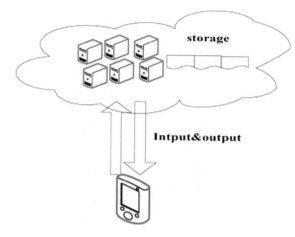

Fig. 5 Agent-client architecture for Mobile Cloud Computing [27]

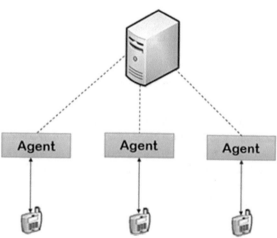

- Agent-client scheme: In this scheme, the cloud gives all assets to cell phones [20]. As is indicated Fig. 5 the cloud creates an operator for every device, which going to fill in as an element to speak with different devices outside the area. Cloudlet a concept that was proposed by Satyanarayanan [21], where the cell phone abuses for virtual machine technology to alter service software to the nearby cloudlet. The mobile device normally works as a meager customer. The cloudlet objective is to decrease execution time [21].
- Collaborated scheme: In this methodology as appeared in Fig. 6 the mobile devices are considered as a piece of the cloud. Subsequently, the aggregate assets are taking by the nearby region. In [22] showed the feasibility of this schema by empowering cell phones inside a network registering structure by executing the BONIC customer on an apple iPhone.

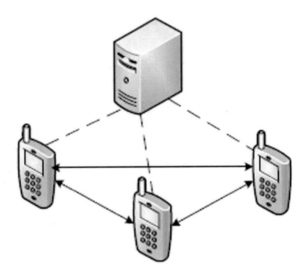

Fig. 6 Collaborated architecture for MCC [27]

3.3 Mobile Edge Cloud

According to ETSI, MEC is defined as follows in [23]: "Mobile edge computing provides an IT service environment and cloud computing capabilities at the edge of the mobile network, within the radio access network (RAN) and in close proximity to mobile subscribers". The Fig. 7 shows a MEC system, in which MEC servers are conveyed in a remote access organize and give processing and capacity storage at the edge of the system.

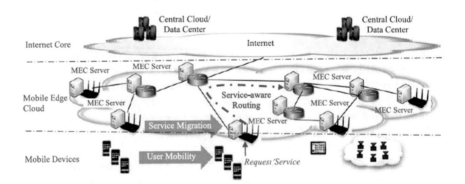

Fig. 7 Mobile edge cloud system overview [51]

3.4 Hadoop

Hadoop is an open source distributed processing framework that issued to distribute processing of large data sets across clusters of computers to run simple programming models [22]. Hadoop is composed of several components that work together to process batch data [24]:

- HDFS: it is the appropriated record framework that organizes capacity and replication over the cluster nodes providing a high-throughput section to application information, it stores intermediate processing results. HDFS consists of two nodes. The first called master or Name Node in which store the location of all files in the local file system, maintains order, and distributes tasks, and the others are slaves called data node it allows to carry out the operations requested by Name Node such as the recovery of the requested blocks, the storage of the data in blocks and it brings to the Name Node a list of the blocks stored.
- MapReduce is Hadoop's local cluster preparing motor. Google presented MapReduce as a programming model to encourage its search processes. it is a matter of breaking down a task into smaller tasks, or more precisely cutting a task involving very large volumes of data into identical tasks relating to subsets of these data. The tasks are then dispatched to different servers, and the results are retrieved and consolidated. MapReduce is based on two steps: Map in which every task is associated to a couple of (keys, value). Then by using the reduce function which associates all the corresponding values to the same key to a single pair (key, value).

3.5 Device to Device

Device to device (D2D) communication is the direct communication between close user devices; they communicate directly without using the central network or the radio communication. In this section, a set of commonly utilized performance metrics for D2D correspondence have been briefly talked about [25].

- System throughput: It is defined as the general throughput of all D2D sets and cell clients inside a cell framework. A higher estimation of throughput means better execution of the calculation. It is estimated in bits/s.
- latency: It is an indicator of the delay among transmission and reception of data. D2D communication for the most part results in lower latency, because of transmission over a small distance.
- Fairness: This is a very critical indicator of evaluating resource allocation for D2D communication.
- The D2D communication can be in two ways:
- Simple caching: if the asked file is stored on another nearby node, the reserving node transmits the document to the asking for node in D2D mode. If the file is not

cached on the local node, the base station transmits the file to the requested node
[23] as shown in Fig. 8.

- Redundant caching: A subset of the nearby node is utilized to transmit parts of
  the record to the downloading node and the first document is recreated at the
  downloading node as shown in Fig. 9.

The operation of this method is quite simple, a new node arrives to the network
(the orange node), asking for a specific file and the Blue nodes store the data, the
green node requests the file and downloads it from the storage nodes (solid arrows),

Fig. 8 D2D simple caching
[31]

Fig. 9 Redundant caching
[31]

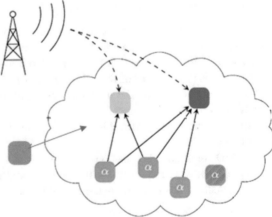

or from the BS (dashed arrow). The repair onto a node (in red) is carried out by transmitting D2D bits from storage nodes (solid arrows) or BS bits from the BS (dashed arrows).

3.6 Software Defined Mobile Edge Cloud

Software defined system (SDSys) is a system that allows the centralized and self-managed control of all IT resources (processing, storage ...). the system architecture is presented in the Fig. 10:

The general architecture is divided into three general layers: the physical (infrastructure) layer, the control (middleware) layer and the applications layer.

The Physical Layer: this layer is a cornerstone since all the other layers are built over it and the different physical resources (storage, security, network, compute, etc.) are combined inside it [26]. The control layer: its divide into two layers the hypervisor layer which responsible to manage and control all devices and the AgNOS layer which responsible of management and administration operations and The application layer: this layer is responsible of all the security applications that are implemented inside [26]. In [26] they suggest a general Software Defined System(SDSys) that supports Mobile Edge Computing (MEC) Fig. 11 illustrates the general framework architecture of the proposed SDMEC System.

The MEC network consists of several domain environments. Each domain is controlled by a local software defined controller which is responsible for keeping smooth communication between the entities inside the same local domain [26]

Fig. 10 The main layers of SDSys architecture [17]

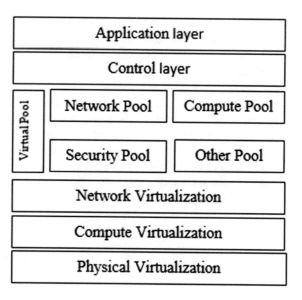

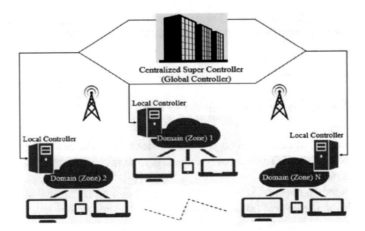

Fig. 11 A prototype reflecting the future architecture of MEC design [17]

- Global Control Layer: is considered the engine of the entire MEC. Made of: SDN controller, SD storage controller unit, security controller unit, SDCompute control unit, SDIoT controller, uni aggregation.
- Local Control Layer: Performs all control procedures and peripheral resource functions found in this layer.

In this article we focus on the SDStorageis one of the subsystems that allows centralized and self-managed control of all IT resources. SDStorage is divided into three main customized components: the host, the switch and the controller (Fig. 12).

- The SDStorage Host: this host stockpiling presents any kind of the capacity exhibits where the client can store his information inside it. It is a virtualized have like any another Host example with some additional parameters related the capacity imitating [27].
- SDStorage Switch: acquires every one of the capacities and parameters of the UserSwitch and executes additional usefulness identified with the SDStore emulation. Inside this switch, a Function Table is made to keep an up-to date data pretty much all the SDStorage Hosts identified with its status and keeps this data accessible to different has in the framework [27].
- SDStorage Controller: Its control several aspects in SDS system, This controller monitors the underlying SDStorage Host and handles different requests by the host [27].

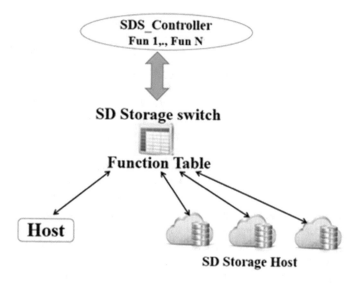

Fig. 12 The general view of the SDStorage experimental framework and its complements [18]

4 Related Works

A lot of research has focused on storage techniques in the mobile network. In this section, we will look at the different works relating with this interest.

We will start with the MCC, mobile cloud computing can be described as a mix of mobile web and cloud computing, with MCC the mobile customer gives provides data processing and storage [28]. The MCC concepts relay on an always-on connectivity to provide high quality mobile access [19]. In [29] they present a different security issues which face the MCC concept, the study divide the security issues into three level the terminal mobile, in which they discuss about the malware and vulnerabilities software. The second issue they speak about the network mobile issue. The network issue is the most dangerous problem that could face the MCC because smartest phone access to the network through Wi-Fi and Bluetooth, etc. So wide access ways will bring greater security threats, for example, the sensitive data leakage or malicious attack. The last issue they discuss is the mobile cloud in which they related the issue in two important things, the first is the platform reliability the cloud platform is powerless to being attacked on account of its high grouping of data assets of users, there are several attack which affect the reliability such as DoS (deny of service) attack which will destroy the platform availability and close the service of the cloud. The second point is data and privacy protection. To solve this security problem several researches have been done. In [30] they present detection system based on detection and prevention cloud AV. The cloud AV is exact example for anti-malware. Cloud AV is another model for malware detection on mobile terminal dependent on giving antivirus as an in-cloud network service. Cloud AV gives a few essential advantages as pursues: better identification of malicious programming; wiping out the effect of antivirus vulnerabilities; upgraded investigation capacities, etc. Also in [31] they present a scientific categorization of current approaches in MCC research on privacy, security and trust. The principle thought is the authentication between weblets that would be conveyed between the cloud and the device. Additionally, in [32] they present a portion of the mobile cloud security structures dependent on authentication and cryptographic tasks. The primary points of interest of MCC is that improve data stockpiling limit and handling power by empowering mobile users to store/get to the expansive data on the cloud. There are many existing stockpiling administrations for mobile devices, for example, Dropbox, Google Drive, etc. [33] with the MCC the mobile user can store substantially more data contrasted than on the PC or devices. The MCC came with several advantages to resolve the storage capacity and the calculation ability limits for the mobile devices, the mobile cloud computing empowers the complex data processing and the enormous data storage executed in the cloud. So the burden of the calculation and storage capacity on mobile device is reduce [29].

The second method is MEC, the mobile edge cloud is a model that incorporates computing and capacity abilities at the edge of the system. MEC has the following characteristics [34]:

- Distributed: the resources of MEC are distributed in various areas, for example, BS, switch as well as the mobile device.
- Bandwidth reduction: since the MEC required less resource to transmit data.
- Interplay with central cloud: the MEC is complementary to the traditional clouds.

The heterogeneity of MEC system makes the conventional trust and authentication instrument insignificant. The principle thought for trust is to know the identity of the element associating with, and the authentication management provides the insurance of trust [35]. In [36], they present the principle issues on the security of the MEC framework among them the infrastructure threads, the MEC exploit diverse technologies, such as, Wi-Fi, LTE, 5G, etc. to build the network, this make the edge infrastructure target to several types of attacks. For instance, DoS attack, the MEC systems are particularly defenseless against distributed DoS attacks, in which some conveyed edge devices that are not well ensured by security protocols can easily be undermined and then used to assault other edge nodes. Another attack is wireless jamming which can easily consume the bandwidth, additionally the men in the middle attack because of the direct transmission between the mobile device and the MEC server which causes interception of the data. In [37] they present a security method to deal with this different attack for the MEC system. In [38] they present a detailed investigation on computing services and storage services. The MEC has a temporary storage because it doesn't support huge data collection but the response time is decreased because of the closeness to the end user.

The third stockpiling technique is Hadoop it a distributed system for parallel processing of data. The advantage of this method it the ability to support different type of data structured and unstructured. In Hadoop there is two important components the MapReduce have two important function the map which responsible of splitting data into a number of data blocks and stores the metadata in a metadata server, and put data blocks into data servers, then the lowest nodes return the results to the parent node that solicited them. This calculates a partial result using the Reduce function. Concerning the security there are a few inquiries about that have been accomplished for the Hadoop security. In [39] they proposed an encryption between name node and data node utilizing the encryption systems (AES), in this mechanism the user must authenticate to have access to the name node. The user has to authenticate himself to access. The user sends a hash function then the name node produces a hash function. these two functions will be compared for purpose of granting access or not. This method keeps up the integrity of data and authentication between nodes, in [40] they proposed a strategy which dependent on Kerberos component so as to build the security of Hadoop or utilizing the bull eye algorithm to deal with the connection between the first and copied data, this algorithm offer access to the data for the approved individual as it were. In [41] they present a Hadoop HBase to build and increase the capacity of storage, and furthermore to improve the read/write functionality in HDFS.

The fourth technique is D2D which mean the direct communication between devices without the need of a Base Station (BS). There are little investigates on this method however concerning the security is too powerless because we take the data locally and we can fall on untrusted device. There are many key issues that faces the D2D communication, which have been recorded beneath include the peer discovery and mode selection, radio resource management, interference management, power control, and security [25]. For the radio resource allocation, they have many algorithms. In [42] they present an Intra cluster D2D retransmission scheme where cooperative relays are adaptively get selected through multicast retransmissions, exploiting the multichannel diversity the objective is to optimize the resource utilization. In [43] they present an admission control performed followed by power allocation to admissible D2D pairs and its cellular user (CU) partners. The objective is to improve spectral efficiency, and overall throughput of the network. Also in [44] present a time division scheduling (TDS) algorithm which is responsible for scheduling the time divided into fixed number of time slots, with a balanced allocation of the D2D pairs in the slots using a location dispersion principle in order to avoid interference. Another approach is presented in [45] for the adaptive time division scheduling algorithm and have as an objective to have a better fairness comparing with the other existing algorithms. There are several attacks faced this technique especially in wireless communication, an attacker can assault the core system, or the user applications, and extract useful data. The D2D is helpless to various security attacks, such as, eavesdropping, data creation, denial of service attack (DoS). To enhance the security of D2D technique they have been some researches. In [46] they use a Diffie–Hellman Secure Key Establishment the objective was to secure key setup between two mobile users. In [47] they use a single hash function, thereby enhancing overall security of the network basing on the encryption. In [48] they use the a beamforming technique, to secure the MIMO D2D communication to provide a high level of secrecy. In [49] they introduce a secure data sharing protocol to Maintain data confidentiality, integrity, provide free-riding resistance and entity authentication. The storage capacity limit is identified with the limit of the gadget itself.

The last method is SDSys. In [33] they proposed a cyber-physical system based on SDSys which have several advantages such as security, scalability and availability. In [26] they proposed another system which is SDMEC, software defined mobile edge cloud which dependent on the standard of the MEC. The MEC network comprises of several domain environments. Every domain is constrained by a local software defined controller which is in charge of keeping smooth correspondence between the elements inside a similar local domain. The system has two layers the control layer which contain different unit Each unit has its roles and functionalities. SDN controller unit which responsible the control and management of network component, the SD storage controller unit to manage the storage resources, all the configurations of the storage resources are generated inside this unit. The ability of SD storage is to facilitate and simplify such complexity. SD Security unit takes the responsibility of monitoring and managing the security aspects of the underlying devices. And the local Controller Layer all the control procedures and functions related to the edge resources reside at this Layer. The advantages of that system that it reduces the response time and it has a huge capacity of storage.

5   Discussion

This article represents a survey of different storage methods for mobile networks. The five methods we treated are as follows: MCC, MEC, Hadoop, D2D and SDMEC. Each of these methods has these strengths as well as weak points.

For the first method, the MCC found that the security paradigm faced challenges, such as the impact of malicious programs by a malicious third party. And as long as science and technology continue to progress, there will be data security issues. Clouds also need a method to improve security. In my opinion, data security issues must include data stored on servers; indeed, the ability to access servers via browsers for internal information can cause multiple attacks. For example, if a hacker attacks a server to steal data, the data stored on the server can be stolen. Security must be improved with authentication, encryption, and digital signature. Also, the security must concern users as well as service providers because a third party may exploit the confidential data. The second problem that face the MCC concept is privacy especially in certain application such as healthcare, since confidential data about a user, his/her own programs and information currently dwell on the cloud (not controlled by the user). So, we can say that the most threads that can face the MCC are: Data lost/stolen devices; Data stolen by mobile malware; Data leakage through poorly written third-party application; Insurance of network access and Vulnerabilities within devices. Another problem is that the data can be sniffed by the intruders during wireless communications. data access can be interrupted because of numerous points. This leads to the data secured with specific administrations. To shield the cell phones from data loss, thin client like anti malware, antivirus ought to be introduced to screen the malicious code. In addition, the response time is long because of the distance between the cloud and the end user. The MCC has its strengths in terms of the storage level, it has a large storage capacity, and users do not care about the infrastructure at the same time as its development. The MCC allow us mobility and supports access to data from anywhere via a web connection. A cloud computing deployment typically relies on a very robust architecture, providing resiliency and redundancy to its users. There is scalability for MCC because it is an integrated feature for cloud deployments.

When we want to talk about the second method, security and resilience are key issues when considering MEC services, such as messaging, navigators, etc., as well as data privacy. With the MEC model, resilience is more difficult because there are interactions with various access technologies, such as Wi-Fi, Bluetooth, LTE, 5G, and so on. To deal with such a problem, we should deploy monitoring mechanisms to become aware of the current level of resiliency of the system/network. To implement the MEC infrastructure we should take on consideration several aspects of trust. Confidentiality: the user information is vulnerable between MEC and cloud communication channel. Interception of the communication will be easy in this case to exploit several data, integrity; the MEC eco system intercorporate with several actors, such as end user, service provider and infrastructure providers, which cause several security challenges such as man in the middle attack who could steal secret information and the availability; the MEC is considered less isolated environment so, it could face a several attack such as DoS attack which cause a non-availability to the system. In addition, the communication delay is reduced due to the proximity between the MEC server and the end user, but even if the respond time is reduced the MEC system has more possibility to be faced to an attack then the other storage methods. The MEC also provides a fast service. And when we talk about storage capacity the MEC has limited capacity because it does not support a huge amount of data. Therefore, the second method needs to be improved at the security level and also at the storage capacity level. in my opinion it is necessary to implement security protocols between the user and the MEC server in order to reinforce the security, resilience and privacy.

As we explained before Hadoop is an open framework that provides a distributed file system (HDFS) and MapReduce. Hadoop is used to store sensitive data, for this reason, the authentication and authorization is necessary to protect data. As we have explained the level of security has been improved with several methods (see Sect. 4), but unfortunately Hadoop always knows privacy and security issues. The most critical security issues that face Hadoop are: Fragmented data: The Hadoop eco system allow multiple copies to ensure redundancy and resilience, this data is available for fermentation and can be shared across multiple servers. As a result, fragmentation is a security issue due to the lack of a security model and node to node communication. Besides the stored data are distributed on the cloud, and to overcome the delay in storing the Hadoop use a public storing. However, this solution leads to several vulnerabilities in transmission and storing. Therefore, there is a need for a security algorithm provides tradeoffs during time delay. In addition the 5 generation will allowed us to download 1 Gbit in 1 s so in one years the number of data generated is 31,104,000 Gbit which is approximately 31 PB and the storage capacity of Hadoop is 25 PB [50]. So, for Hadoop we should increase the storage capacity to deal with the growth of data. Beside the HDFS make a copy for each data, so the capacity of storage will be exhausted rapidly. For this reason, the necessity to improve the security as well as the storage capacity is a critical point to deal with such the data generated. For this reason, we should reinforce the security for Hadoop by: Analyzing the environment specially when we talk about mobile network; classify the data to detect the data which should be stored or deleted and implement a designed security model.

D2D has less security comparing with the other method, because the data is stored locally in the device. Once the mobile device is damaged or lost, the valuable information treasured in the device is lost altogether. So, if the cloud storage can be integrated with D2D communication for periodical data backup of a mobile client, the risk of data lost can be minimized. Also, with this implementation we could increase the storage capacity of D2D, because the capacity is related to the device, so we can say that the capacity is limited. In addition, the security in this method is too weak, because there may be a malicious mobile device in the local area. For this reason, to enhance the security we should implement a mechanism to identify the user equipment (UE), with the use of a cryptographic techniques. Concerning the respond time, the D2D have the less respond time because of the data is loaded and stored in the local device or even in a base station in the local area.

The SDMEC is a storage method which guaranties a huge capacity of storage. The SDMEC combine between the functionality of SDN and the MEC system to benefit from the advantages of both systems. In my opinion the idea behind this though is to enhance the capacity of the edge network concerning the storage capacity. The SDMEC based on two layers, the control layer which contain different resources such as SD Security which takes the responsibility of checking and managing the security aspects of the underlying devices and hosts. To maintain the security on the system, we should also implement a strategy of cryptographic technique to enhance the security concerning this system.The SD Storage; this unit is responsible of controlling and managing the storage resources in the system such as local databases and cloud storage. This unit increase the capacity storage on the level of edge. Then the local domain which responsible of all the control procedures and functions related to the edge resources resides at this layer. Also, with this system such as: more control, flexibility it will enable innovation with the implementation of the edge, adaptability it could take on consideration much more devices and gadgets, low cost and interpretability. The implementation of such architecture layering system has numerous advantages for example, reducing the delay; when the requests are served locally, the system performance is enhanced by reducing the delay. Beside it's considered as a scalable it integrating with several local controllers to facilitate the process of expansion of the network.

6   Conclusion

In this paper we present an overview of the different evolution of mobile communications through all its generations. From the first generation which based on analogic FM to an IP-based data network. We showed that with the advent of 5G, more data has been generated and the importance of data storage is becoming critical. In this paper, we concisely reviewed advantages and disadvantages of several stockpiling methods, and analyzes those methods basing on the most critical criteria: the stockpiling limit, level of security and the response time. Then, according to the issues we gave the current approaches to deal with the problem itself.

References

1. Elhoseny, M., Abdelaziz, A., Salama, A., Riad, A.M., Sangaiah, A.K., Muhammad, K.: A hybrid model of internet of things and cloud computing to manage big data in health services applications. Futur. Gener. Comput. **86**, 1383–1394 (2018)
2. Elhoseny, M., Shankar, K., Lakshmanaprabu, S.K., Maseleno, A., Arunkumar, N.: Hybrid optimization with cryptography encryption for medical image security in Internet of Things. Neural Comput. Appl. 1–15 (2018). https://doi.org/10.1007/s00521-018-3801-x
3. Hurrah, N.N., Parah, S.A., Loan, N.A., Sheikh, J.A., Elhoseny, M., Muhammad, K.: Dual watermarking framework for privacy protection and content authentication of multimedia. Futur. Gener. Comput. Syst. **94**, 654–667 (2019)
4. Muhammad, K., Khan, S., Elhoseny, M., Ahmed, S.H., Baik, S.W.: Efficient fire detection for uncertain surveillance environment. IEEE Trans. Ind. Inform. (2019)
5. Boveiri, H.R., Elhoseny, M.: A-COA: an adaptive cuckoo optimization algorithm for continuous and combinatorial optimization. Neural Comput. Appl. DOI:https://doi.org/10.1007/s00521-018-3928-9 (First Online: 15 December 2018)
6. Rao, H., Shi, X., Rodrigue, A.K., Feng, J., Xia, Y., Elhoseny, M., Yuan, X., Gu, L.: Feature selection based on artificial bee colony and gradient boosting decision tree. Appl. Soft Comput. DOI:https://doi.org/10.1016/j.asoc.2018.10.036 (Online: 5 November 2018)
7. Rappaport, T.S., et al.: Millimeter wave mobile communications for 5G cellular: it will work! IEEE Access **1**, 335–349 (2013)
8. Myung, H.G., Goodman, D.J.: Single Carrier FDMA: A New Air Interface for Long Term Evolution. Wiley, Chichester, UK (2008)
9. Jindal, N., Goldsmith, A.: Dirty-paper coding versus TDMA for MIMO broadcast channels. IEEE Trans. Inf. Theory **51**(5), 1783–1794 (2005)
10. G PPP Architecture Working Group view on 5G Architecture. WP-July-2016
11. GSMA, 5G, the Internet of Things (IoT) and Wearable Devices. September-2017
12. Jungnickel, V., et al.: The role of small cells, coordinated multipoint, and massive MIMO in 5G. IEEE Commun. Mag. **52**(5), 44–51 (2014)
13. mobicom10-duplex.pdf
14. Armbrust, M., et al.: A view of cloud computing. Commun. ACM **53**(4), 50 (2010)
15. Jadeja, Y., Modi, K.: Cloud computing - concepts, architecture and challenges. In: 2012 International Conference on Computing, Electronics and Electrical Technologies (ICCEET), Nagercoil, Tamil Nadu, India, pp. 877–880 (2012)
16. Rimal, B.P., Choi, E., Lumb, I.: A taxonomy and survey of cloud computing systems. In: 2009 Fifth International Joint Conference on INC, IMS and IDC, Seoul, South Korea, pp. 44–51 (2009)

17. Mathur, P., Nishchal, N.: Cloud computing: new challenge to the entire computer industry. In: 2010 First International Conference On Parallel, Distributed and Grid Computing (PDGC 2010), Solan, India, pp. 223–228 (2010)
18. Mathur, P., Nishchal, N.: Cloud computing: new challenge to the entire computer industry (:unav), october 2010 (2010)
19. Guan, L., Ke, X., Song, M., Song, J.: A survey of research on mobile cloud computing. In: 2011 10th IEEE/ACIS International Conference on Computer and Information Science, Sanya, China, pp. 387–392 (2011)
20. Satyanarayanan, M., Bahl, P., Caceres, R., Davies, N.: The case for VM-based cloudlets in mobile computing. IEEE Pervasive Comput. **8**(4), 14–23 (2009)
21. Black, M., Edgar, W.: Exploring mobile devices as Grid resources: using an x86 virtual machine to run BOINC on an iPhone. In 2009 10th IEEE/ACM International Conference on Grid Computing, pp. 9–16 (2009)
22. Ibtissame, K., Yassine, R., Habiba, C.: Real time processing technologies in big data: comparative study. In 2017 IEEE International Conference on Power, Control, Signals and Instrumentation Engineering (ICPCSI), Chennai, pp. 256–262 (2017)
23. Paakkonen, J., Hollanti, C., Tirkkonen, O.: Device-to-device data storage for mobile cellular systems. In: 2013 IEEE Globecom Workshops (GC Wkshps), Atlanta, GA, pp. 671–676 (2013)
24. Sharma, P.P., Navdeti, C.P.: Securing big data hadoop: a review of security issues, threats and solution. Int. J. Comput. Sci. Inf. Technol. **5**, 6 (2014)
25. Gandotra, P., Kumar Jha, R., Jain, S., A survey on device-to-device (D2D) communication: architecture and security issues. J. Netw. Comput. Appl. **78**, 9–29 (2017)
26. Jararweh, Y., Doulat, A., Darabseh, A., Alsmirat, M., Al-Ayyoub, M, Benkhelifa, E.: SDMEC: software defined system for mobile edge computing. In: 2016 IEEE International Conference on Cloud Engineering Workshop (IC2EW), Berlin, Germany, pp. 88–93 (2016)
27. Darabseh, A., Al-Ayyoub, M., Jararweh, Y., Benkhelifa, E., Vouk, M., Rindos, A.: SDStorage: a software defined storage experimental framework. In: 2015 IEEE International Conference on Cloud Engineering, Tempe, AZ, USA, pp. 341–346 (2015)
28. Christensen, J.H.: Using RESTful web-services and cloud computing to create next generation mobile applications. In: Proceedings of the 24th ACM SIGPLAN Conference Companion on Object Oriented Programming Systems Languages and Applications, New York, NY, USA, pp. 627–634 (2009)
29. Suo, H., Liu, Z., Wan, J., Zhou, K.: Security and privacy in mobile cloud computing. In: 2013 9th International Wireless Communications and Mobile Computing Conference (IWCMC), Sardinia, Italy, pp. 655–659 (2013)
30. CloudAV: N-Version Antivirus in the Network Cloud
31. Fernando, N., Loke, S.W., Rahayu,W.: Mobile cloud computing: a survey. Futur. Gener. Comput. Syst. **29**(1), 84–106 (2013)
32. Khan, A.N., Mat Kiah, M.L., Khan, S.U., Madani, S.A.: Towards secure mobile cloud computing: a survey. Futur. Gener. Comput. Syst. **29**(5), 1278–1299 (2013)
33. Darabseh, A., Freris, N.M.: A software defined architecture for cyberphysical systems. In: 2017 Fourth International Conference on Software Defined Systems (SDS), Valencia, Spain, pp. 54–60 (2017)
34. Burt, J.: Fog computing aims to reduce processing burden of cloud systems. eWeek (2010)
35. Mobile edge computing, Fog et al.: A survey and analysis of security threats and challenges - ScienceDirect
36. Shirazi, S.N., Gouglidis, A., Farshad, A., Hutchison, D.: The extended cloud: review and analysis of mobile edge computing and fog from a security and resilience perspective. IEEE J. Sel. Areas Commun. **35**(11), 2586–2595 (2017)
37. Xiao, L., Wan, X., Dai, C., Du, X., Chen, X., Guizani, M.: Security in mobile edge caching with reinforcement learning. IEEE Wirel. Commun. **25**(3), 116–122 (2018)
38. El-Sayed, H., et al.: Edge of things: the big picture on the integration of edge, IoT and the cloud in a distributed computing environment. IEEE Access **6**, 1706–1717 (2018)

39. Abouelmehdi, K., Beni-Hssane, A., Khaloufi, H., Saadi, M.: Big data emerging issues: hadoop security and privacy. In: 2016 5th International Conference on Multimedia Computing and Systems (ICMCS), Marrakech, Morocco, pp. 731–736 (2016)
40. Adluru, P., Datla, S.S., Zhang, X.: Hadoop eco system for big data security and privacy. In: 2015 Long Island Systems, Applications and Technology, Farmingdale, NY, USA, pp. 1–6 (2015)
41. Zhang, D.-W., Sun, F.-Q., Cheng, X., Liu, C.: Research on hadoop-based enterprise file cloud storage system. In: 2011 3rd International Conference on Awareness Science and Technology (iCAST), Dalian, China, pp. 434–437 (2011)
42. Zhou, B., Hu, H., Huang, S.-Q., Chen, H.-H.: Intracluster device-to-device relay algorithm with optimal resource utilization. IEEE Trans. Veh. Technol. 62(5), 2315–2326 (2013)
43. Feng, D., Lu, L., Yuan-Wu, Y., Li, G.Y., Feng, G., Li, S.: Device-to-device communications underlaying cellular networks. IEEE Trans. Commun. 61(8), 3541–3551 (2013)
44. Chen, B., Zheng, J., Zhang, Y.: A time division scheduling resource allocation algorithm for D2D communication in cellular networks. In: 2015 IEEE International Conference on Communications (ICC), London, pp. 5422–5428 (2015)
45. Zheng, J., Chen, B., Zhang, Y.: An adaptive time division scheduling based resource allocation algorithm for D2D communication underlaying cellular networks, (:unav), december 2014 (2014)
46. Shen, W., Hong, W., Cao, X., Yin, B., Shila, D.M., Cheng, Y.: Secure key establishment for device-to-device communications. In: 2014 IEEE Global Communications Conference, Austin, TX, USA, pp. 336–340 (2014)
47. Abd-Elrahman, E., Ibn-khedher, H., Afifi, H., Toukabri, T.: Fast group discovery and non-repudiation in D2D communications using IBE. In: 2015 International Wireless Communications and Mobile Computing Conference (IWCMC), Dubrovnik, Croatia, pp. 616–621 (2015)
48. Jayasinghe, K., Jayasinghe, P., Rajatheva, N., Latva-aho, M.: Physical layer security for relay assisted MIMO D2D communication. In: 2015 IEEE International Conference on Communication Workshop (ICCW), London, United Kingdom, pp. 651–656 (2015)
49. Zhang, A., Chen, J., Hu, R.Q., Qian, Y.: SeDS: secure data sharing strategy for D2D communication in LTE-advanced networks. IEEE Trans. Veh. Technol. 65(4), 2659–2672 (2016)
50. Shvachko, K., Kuang, H., Radia, S., Chansler, R.: The hadoop distributed file system. In: 2010 IEEE 26th Symposium on Mass Storage Systems and Technologies (MSST), Incline Village, NV, USA, pp. 1–10 (2010)
51. Liu, H., Eldarrat, F., Alqahtani, H., Reznik, A., de Foy, X., Zhang, Y.: Mobile edge cloud system: architectures, challenges, and approaches. IEEE Syst. J. 12(3), 2495–2508 (2018)

Application of Artificial Intelligence Approach for Optimizing Management of Road Traffic

Charlène Béatrice Bridge-Nduwimana, Abdessamad Malaoui and Jilali Antari

Abstract An approach based on the artificial intelligence is proposed for the management of road traffic. By a fuzzy system, we are looking for purely numerical parametric characteristics and those that influence its structure. In fact, we use input and output data from a portion of the road traffic to identify a fuzzy model which makes possible the evaluation of the results of the estimated parameters obtained. This has been achievable through the combination of parametric and structural adjustment algorithms with the backpropagation algorithm. Consequently, the obtained results show that adaptive models are successfully used in the analysis and the management of road traffic through the efficiency of this combination.

Keywords Road traffic · Fuzzy system · Artificial intelligence · Combination

1 Introduction

Overall, we will deal with the problem of identifying fuzzy models from input-output data. Sometimes we need to have the parameters of a system without knowing all the members. Through an example we demonstrate this fuzzy identification [1]. With the fuzzy modeling formalism of systems focusing particularly on the Takagi-Sugeno model [2], we represent the nonlinear behavior of a system by a composition of "If … Then" rules, concatenating a set of sub-models locally linear. In what follows, the fuzzy model has been identified for a signal based on data in autoregressive way. This method is simple and allowed us to generate data without other variables in

C. B. Bridge-Nduwimana (✉) · A. Malaoui
Polydisciplinary Faculty of BeniMellal, Sultan Moulay Slimane University, Beni-Mellal, Morocco
e-mail: cbridgebeatrice@gmail.com

A. Malaoui
e-mail: a.malaoui@usms.ma

J. Antari
Polydisciplinary Faculty of Taroudant, Ibn Zohr University, Agadir, Morocco
e-mail: j.antari@uiz.ac.ma

© Springer Nature Switzerland AG 2020
M. Elhoseny and A. E. Hassanien (eds.), *Emerging Technologies for Connected Internet of Vehicles and Intelligent Transportation System Networks*, Studies in Systems, Decision and Control 242, https://doi.org/10.1007/978-3-030-22773-9_5

61

addition. The model is capable to predict other data on the process being studied once the optimization phase is over.

To build such models, we approach the application of a competitive agglomeration method: the algorithm of Gustafson and Kessel [3, 4] which belongs to fuzzy clustering methods based on the minimization of an objective function. Finally, after having considered the general methodology for the construction of the Takagi-Sugeno fuzzy model from data, we will comment on what we will obtain as results.

This paper is organized as follows. Section 2 develops Takagi-Sugeno's fuzzy model and discusses the types of adjustments made to fuzzy systems. Section 3 concisely discusses about achievements in scientific research related to the subject. And Sect. 4 explains the approaches adopted and the results of the simulations are established in Sect. 5. Finally, we conclude this paper in Sect. 6.

For all our simulations, we carried out fuzzy clustering simulations for the structural adjustment, parametric adjustment simulations with the GLS and WLS algorithms [5] combined with the backpropagation [6] algorithm simultaneously. They give Root Mean Square Error (RMSE) on the validation set between the predicted values and measured data.

2 Takagi-Sugeno's (TS) Fuzzy Models

Developing an artificial intelligence to process a large amount of data is therefore an excellent idea. Today, the practical advantages of artificial intelligence are in fairly pragmatic operations. A fuzzy system is a system that integrates human expertise and aims to emulate the reasoning of human experts in complex systems. It is an important part of artificial intelligence.

In fuzzy systems, the basic idea is to model processes as would the human being [7]. The relationships between input and output variables are explicitly represented in the form of "If … Then …" rules, i.e.: If HYPOTHESIS (antecedent part) Then CONCLUSION (consequent part) [8]. The above form allows to interpret the results and to determine the action of each rule and express an inference [9] mechanism such that if a fact (hypothesis) is known, then another fact (conclusion) can be inferred. The Takagi-Sugeno fuzzy model [1] uses linear functions in the consequent part: for i rule and j output, we have: $y_j^i = f_j^i(x)$.

So, it can be seen as a combination [10] of the linguistic model and the mathematical regression [10] model in the sense that the antecedents describe fuzzy regions in the input space in which the consequent functions are valid. Basically, this model can encode the expertise, either directly from the prior knowledge of the problem, or indirectly from a set of learning data. It is very easy to identify because the conclusion of each rule is linear and its parameters can be estimated from the numerical data using optimization [9] methods such as least squares algorithms.

We will use more particularly these models because they allow to approach non-linear systems by a combination [10] of several linear and simple local models. They are written as follows [11]:

$$R_i : If \ x_t \ is \ A_i \ Then \ \hat{y}_{t,i} = \beta_{0i} + x_t^T \beta_i \tag{1}$$

$R_i (i = 1, 2, \ldots, c)$ indicates the ith fuzzy rule, $x_t (t = 1, 2, , N)$ is the input variable $(x_t \in R^n)$, $\hat{y}_{t,i}$ is the output of the ith rule relative to the input x_t, A_i a fuzzy set [12] and $\beta_i = (\beta_1, \beta_2, \ldots, \beta_n)^T$.

2.1 Structural Adjustment

Structural adjustment is about determining the correct number of rules to use in a fuzzy system [13, 14]. The structure to be searched for will have to be rich enough to allow for optimal learning, but not too much to avoid noise modeling in the data.

2.2 Parametric Adjustment

Once the number of rules determined, it is necessary to estimate the parameters (β_i) for each conclusion of the rules. If we have:

$$W_i = \begin{pmatrix} \mu_{i1} & 0 & \cdots & 0 \\ 0 & \mu_{i2} & \cdots & 0 \\ \vdots & \vdots & \ddots & \vdots \\ 0 & 0 & \cdots & \mu_{iN} \end{pmatrix}, \quad X = \begin{pmatrix} x_1 \\ \vdots \\ x_N \end{pmatrix}, \quad y = \begin{pmatrix} y_1 \\ \vdots \\ y_N \end{pmatrix} \tag{2}$$

And if we have:

$$X_e = \begin{bmatrix} 1 & X \end{bmatrix}, \quad \tilde{X} = \begin{bmatrix} W_1 X_e & W_2 X_e & \cdots & W_c X_e \end{bmatrix} \tag{3}$$

2.2.1 Weighted Least Square (WLS)

The localized linearization method causes the resolution of c independent optimization problems. Linear parameters obtained do not depend on how the rules are aggregated. The criterion to be minimized is:

$$J = \sum_{i=1}^{c} \sum_{t=1}^{N} w_i \left(\hat{y}_t - x_t^T \beta_i \right)^2 \tag{4}$$

The determination of the linear parameters β_i passes by the minimization of the criteria of each local model. This amounts to solving c independent weighted least squares problems whose solution is:

$$\beta_i = \left(X_e^T W_i X_e \right)^{-1} X_e^T W_i y \tag{5}$$

2.2.2 Global Least Square (GLS)

The global system resulting from this method [11], approximates the database with more perfection. But nothing tells us that the linear models thus obtained are optimal in their areas of expertise. Linearparameters are obtained by solving the equation:

$$\tilde{X}\beta = y \ or \ \left(\tilde{X}^T \tilde{X} \right) \beta = \tilde{X}^T y \tag{6}$$

The criterion to be minimized for GLS is:

$$J = \frac{1}{2} \sum_{t=1}^{N} \left(\hat{y}_t - y_t \right)^2 \tag{7}$$

3   A State of the Art

Accurate data from accident and road databases can be essential for modeling, mapping, identifying hazardous road segments and other studies to make decisions in a road network. Researches relating to the problems that are in the databases of road traffic and which propose solutions exist and bring a plus. Among them, we have:

C. Yixin and X. Deyun [2] represent an extension of the Takagi-Sugeno-Kang model (ETSK). Its analytic expression has been delivered and an algorithm to identify such a model has been proposed. TSK with variable weight (VWTSK) was made to present the fuzzy controller algorithm of the ETSK model even definition of fuzzy rules since they are roughly equivalent. The simulation of this algorithm shows that the ETSK model can give more precision on the long-term predictions and the control algorithm can reach a better control more efficient that Proportional-Integral-Derivative fuzzy regulation (PID). Furthermore, an adaptive control identification or method for a system based on FCM-KNN (Fast Fuzzy C-Means—K-Nearest Neighbors) and PSO (Particle Swarm Optimization) has been proposed by Rastegar et al. [1]. The model identify the structure and parameters of the nonlinear model: the fuzzy set and the number of rules, and the location of the membership functions are automatically pulled from the system data. In comparison with other identification methods, larger values corresponding to a lower number of fuzzy rules have been achieved. Thus, their results showed that the proposed control model can control the process just by using a database of the TS Adaptive Fuzzy Model initialization process. On the other hand, C. N. Babu and B. E. Reddy [15], for the prediction of time-based internet traffic that is very volatile in nature, have explored the applicability of various forecasting models. They considered during their study ARIMA (AutoRegressive Integrated Moving Average), ANN (Artificial Neural Network), Zhang's ARIMA-ANN hybrid model, Khashei and Bijari's ARIMA-ANN hybrid model, the ARIMA-ANN multiplicative model, MA (Moving Average Filter) filter based on the ARIMA-ANN hybrid model. One-step and/or multi-step predictions have been made. The measures of the error performance, MAE (Mean Absolute Error) and RMSE (Root Mean Squared Error) are used to evaluate accuracy [16–18]. The results of the forecast in both cases showed that the MA filter based on the ARIMA-ANN hybrid model outperformed all the others models, both in terms of MAE and RMSE and is therefore suitable for more accurate prediction of internet traffic data.

4 Approaches

At first, we generate [19] fuzzy rules of the TS model. We use classification [14, 20–23] algorithm (Gustafson Kessel: GK) [24] to estimate the number and initial positions of cluster centers where each one allows us to determine a fuzzy relationship between inputs and outputs by checking their similarity. Then we adopt fuzzy generation algorithm to predict fuzzy output. Even though there is no indication of

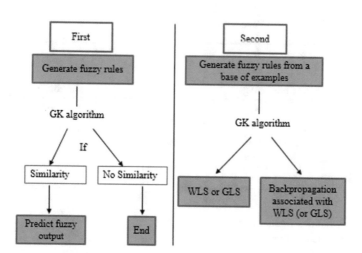

Fig. 1 Main Approaches

this kind of problem, the GK algorithm [4] makes it possible by giving the state or the quality of the road traffic taken as example from the output of the fuzzy model.

Secondly, we generate fuzzy rules from a base of examples where we want to classify the outputs into a set of predefined fuzzy classes. Then we use the first approach for fuzzy rule generation, we apply a weighted or generalized least squares fit to compare the fuzzy and predefined outputs of our model (Fig. 1).

5   Results

5.1   First Simulation

In Figs. 2 and 3, we represent the results after a simulation realized on matlab R2017a on a computer of properties: Intel Core i5 processor with 2.60 GHz and 8 GB of Ram. There are the membership functions, the linear β_i and nonlinear parameters that are estimated with the WLS or GLS algorithms.

And in the following tables, Tables 1 and 2, we present the results on different values of the set of input in order to capture the sensitivity and the effects of these two methods on our example.

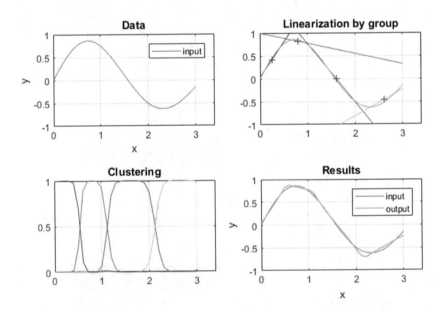

Fig. 2 For 31 values with WLS algorithm

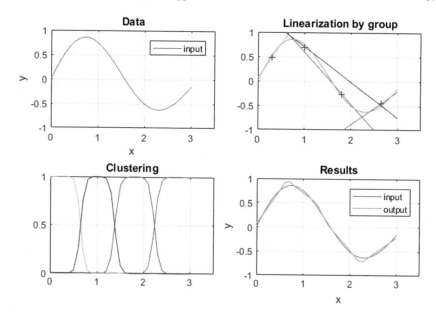

Fig. 3 For 31 values with GLS algorithm

Table 1 Results for WLS algorithm

Values	Clusters	RMSE	Errors
11	4	0.0076	6.3315×10^{-4}
21	4	0.0217	0.0099
31	4	0.0467	0.0676
51	4	0.0580	0.1717
71	6	0.0498	0.1763
151	8	0.0494	0.3678
501	26	0.0226	0.2554

Table 2 Results for GLS algorithm

Values	Clusters	RMSE	Errors
11	4	0.0074	6.0077×10^{-4}
21	4	0.0213	0.0095
31	4	0.0460	0.0655
51	4	0.0582	0.1730
71	6	0.0498	0.1758
151	8	0.0446	0.2997
501	28	0.0185	0.1709

Table 3 Results for WLS algorithm

Class	Clusters	RMSE	Errors
CL1	4	0.0324	0.0326
CL2	4	1.3714×10^{-4}	5.8305×10^{-7}
CL3	4	5.3609×10^{-4}	8.9092×10^{-6}

Table 4 Results for GLS algorithm

Class	Clusters	RMSE	Errors
CL1	4	0.0313	0.0304
CL2	4	1.3724×10^{-4}	5.8389×10^{-7}
CL3	4	5.3909×10^{-4}	9.0088×10^{-6}

Table 5 Results for WLS and backpropagation algorithm

Class	Clusters	RMSE	Errors
CL1	4	0.0311	0.0299
CL2	4	1.3712×10^{-4}	5.8284×10^{-7}
CL3	4	5.3922×10^{-4}	9.0137×10^{-6}

Table 6 Results for GLS and backpropagation algorithm

Class	Clusters	RMSE	Errors
CL1	4	0.0298	0.0276
CL2	4	1.3723×10^{-4}	5.8380×10^{-7}
CL3	4	5.3864×10^{-4}	8.9941×10^{-6}

5.2 Second Simulation

In order to estimate road traffic [25] parameters, we use a database consisting of daily measured values for January 2012. These values were taken in the Gironde region, a french department located in the south-west of the country in the New Aquitaine region. The performance criterion chosen remains the root mean squared error. We chose to sort three complete classes of the database that normally contains several classes (see Tables 3 and 4). We want to estimate the ratio between the length and speed provided. This is indeed an important data that characterizes the portion of the road taken into consideration.

A better approximation is an added value for prediction which is very useful in applications because it allows to generate traffic data for localities where measurements are not available. When the linear function is bounded and the activation function is derivable, it is possible to use powerful learning algorithms based on the search for a minimum of the error function, in particular the backpropagation [26] of the gradient which includes hidden layers. So, to update the connection weight within a network so that it succeeds in the task that is asked of it, and thus apply our example to artificial intelligence [27], we used the method of backpropagation [28, 6] (see Tables 5 and 6).

5.3 Comments

The results obtained in Figs. 1 and 2 show that whatever the WLS or GLS algorithm we can have a good estimate justified by the mean squared error not by the graph. We can note that the choice of the number of clusters in the algorithm depends on how much data appears to us. It is important to find a number of clusters that best estimate the input. In statistics, generalizedleast squares are a technique for estimating parameters unknown in a linear regression model. Indeed, GLS is used to perform a linear regression when thereis some degree of correlation between values in a model. In this case, the classical weighted least square method can be statistically ineffective or even give misleading inferences.In accordance with previous results (in Tables 5 and 6), the results show that backpropagation associated with GLS, under conditions of poor specification, provides realistic indices of model implementation and less biased parameter values for paths that overlap with the real model. However, despite the recommendations of the literature that WLS should be used when data is not distributed normally, we find that under no circumstances is the WLS method better than the other two methods of estimating parameters in terms of bias and implementation. In fact, only for large sample and for implementation indices close to those obtained for backpropagation and the GLS method. In addition for wrongly specified models, WLS gives low estimates reliable and overly optimistic values of fit. With simulations performed with WLS/GLS methods associated with backpropagation simultaneously, if we increase the number of iterations there is noise added because it takes more time during the simulation. It will be the same if we increase the number of hidden layers where a certain amount of information will be lost. It is advisable to consider few layers hidden to avoid noise and a number of reasonable iterations that best justifies the convergence of the error towards zero. Then the class CL1 is the best estimated by the GLS associated with backpropagation method which is more confident and is likely to help us make a decision.

6 Conclusion

In this work the fuzzy model TS has been identified for our example and for a signal of the road traffic studied based on data in an autoregressive way. This method is simple and allowed us to generate data without the need to use other variables in addition. Moreover, once the optimization phase is over, the model is capable to predict other data on the process being studied. We used design methods based on a learning that allows to iteratively define the best set of parameters: the optimization of fuzzy rules (WLS, GLS) and the optimization of membership functions. We also have proposed one of the WLS or GLS optimization models with backpropagation to test the convergence of the error. The results obtained show that even with a non linear we can hope to obtain quite satisfactory performance.

References

1. Rastegar, S., et al.: Self-adaptative Takagi-Sugeno model identification methodology for industrial control processes (2011)
2. Chen, Y., Xiao, D.: Fuzzy identification and control algorithms based on an ETSK model. **32**, 5456–5461 (1999)
3. Palacio, V.G.: Modélisation et commande floues de types TAKAGI-SUGENO appliquées à un bioprocédé de traitement des eaux usées. Ph.D. thesis (2007)
4. Wu, H., et al.: Bisimulations for fuzzy transition systems revisited. **99**, 1–11 (2018)
5. Olsson U.H. et al.: The performance of ML, GLS, and WLS estimation in structural equation modeling under conditions of misspecification and nonnormality. 557–595 (2000)
6. Elmzabi, A.: Une approche adaptative basée sur la clustering flou: Outil d'aide à la gestion intelligente du réseau. Ph.D.thesis (2005)
7. Bouzid, H.E., Benmeriem, S.: Application de la technique de la logique floue pour la prédiction de l'amorçage des intervalles d'air pointes-plans (2013)
8. Ishibushi, H., et al.: Selecting fuzzy if-then rules for classification problems using genetic algorithms. **3**, 260–270 (1995)
9. Milan, S.G. et al.: Fuzzy optimization model and fuzzy inference system for conjunctive use of surface and groundwater resources. **566**, 421–434 (2018)
10. Tomar, A.S., et al.: Traffic management using logistic regression with fuzzy logic. 451–460 (2018)
11. Iqdour, R.: Modélisation des séries temporelles par les systèmes flous et les réseaux de neurones: application à la prédiction des processus météorologiques. Ph.D. thesis (2006)
12. Yang, B., Hu, B.Q.: Communication between fuzzy information systems using fuzzy covering-based rough sets. **103**, 414–436 (2018)
13. Dass, A., Srivastava, S.: Identification and control of dynamical systems using different architectures of recurrent fuzzy system (2018)
14. Kashani, A.T., Mohaymany, A.S.: Analysis of the traffic injury severity on two-lane, two-way rural roads based on classification tree models. **49**, 1314–1320 (2011)
15. Babu, C.N., Reddy, B.E.: Performance comparison of four new ARIMA-ANN prediction models on internet traffic data. 1, 67–75 (2015)
16. Boveiri, H.R., Elhoseny, M.: A-COA: an adaptive cuckoo optimization algorithm for continuous and combinatorial optimization. Neural Comput. Appl., first online: 15 Dec 2018. DOI:https://doi.org/10.1007/s00521-018-3928-9
17. Haidi Rao, Xianzhang Shi, Ahoussou Kouassi Rodrigue, Juanjuan Feng, Yingchun Xia, Mohamed Elhoseny, Xiaohui Yuan, Lichuan Gu, Feature selection based on artificial bee colony and gradient boosting decision tree, Applied Soft Computing, Available online 5 November 2018. DOI:https://doi.org/10.1016/j.asoc.2018.10.036
18. Chen, Tian, Sihang, J., Yuan, X., Elhoseny, M., Ren, F., Fan, M., Chen, Z.: Emotion recognition using empirical mode decomposition and approximation entropy. Comput. Electr. Eng. **72**, 383–392 (2018)
19. Ishibushi, H., et al.: A simple but powerful heuristic method for generating fuzzy rules from medical data. 251–270 (1997)
20. Murugan, B.S., Elhoseny, M., Shankar, K., Uthayakumar, J.: Region-based scalable smart system for anomaly detection in pedestrian walkways. Comput. Electr. Eng. **75**, 146–160, (2019)
21. Shankar, K., Elhoseny, M., Lakshmanaprabu, S.K., Ilayaraja, M., Vidhyavathi, R.M., Alkhambashi, M.: Optimal feature level fusion based ANFIS classifier for brain MRI image classification. Concurr. Comput.: Pract. Exp. (2018)
22. Muhammad, K., Khan, S., Elhoseny, M., Ahmed, S.H., Baik, S.W.: Efficient fire detection for uncertain surveillance environment. IEEE Trans. Ind. Inform. (2019)
23. Hurrah, N.N., Parah, S.A., Loan, N.A., Sheikh, J.A., Elhoseny, M., Muhammad, K.: Dual watermarking framework for privacy protection and content authentication of multimedia. Futur. Gener. Comput. Syst. **94**, 654–667 (2019)

24. Babuska, R., et al.: Improved covariance estimation Gustafson-Kessel clustering. 1081–1085 (2002)
25. Ihueze, C., Onwurah, U.: Road traffic accidents prediction modelling: an analysis of Anambra state, Nigeria. 21–29 (2018)
26. Wang, L., Mendel, J.: Back-propagation fuzzy system as nonlinear dynamic system identifiers. In: Conference (1992)
27. Johnson, K.W., et al.: Artificial intelligence in cardiology. **71**, 2668–2679 (2018)
28. Hassabis, D., et al.: Neuroscience-inspired artificial intelligence. 245–258 (2017)

Mobility Condition to Study Performance of MANET Routing Protocols

Hind Ziani, Nourddine Enneya, Jihane Alami Chentoufi
and Jalal Laassiri

Abstract The performance of a Mobile Ad hoc Network (MANET) is closely related to the capability of routing protocols to adapt themselves to unpredictable changes of topology network and link status. In simulations, performance studies of routing protocols depend on the chosen mobility model. Consequently, the comparison of the obtained performance results becomes more difficult even it runs on the same simulation environment. To solve this problem it is necessary add a mobility condition, independently, of the used mobility models. In this paper, we define this mobility condition and show how it makes performance comparaison of routing protocols judicious and easier.

Keywords Mobility · Node mobility · Network performance · QoS · Routing protocols · Ad hoc networks · MANETs · Manet routing protocols

1 Introduction

A mobile ad hoc network (MANET) [1] is a collection of mobile nodes that cooperatively communicate with each other without any pre-established infrastructure such as a centralized access point. These nodes may be computers or devices such as laptops, PDAs, mobile phones and pocket PCs which have in common a wireless connectivity. The idea of forming a network with no infrastructure originates from

H. Ziani (✉) · N. Enneya · J. A. Chentoufi · J. Laassiri
Informatics, Systems and Optimization Laboratory, Faculty of Sciences,
University of Ibn Tofail, BP 133, Kenitra, Morocco
e-mail: hind.ziani@uit.ac.ma

N. Enneya
e-mail: enneya@uit.ac.ma

J. A. Chentoufi
e-mail: j.alami@uit.ac.ma

J. Laassiri
e-mail: laassiri@uit.ac.ma

© Springer Nature Switzerland AG 2020
M. Elhoseny and A. E. Hassanien (eds.), *Emerging Technologies for Connected
Internet of Vehicles and Intelligent Transportation System Networks*, Studies in Systems,
Decision and Control 242, https://doi.org/10.1007/978-3-030-22773-9_6

DARPA (Defence Advanced Research Projects Agency) packet radio network's days [2]. Due to the fact that nodes change their physical location by moving around, the network topology may unpredictably change, which causes changes in link status between each node and its neighbors. Thus, nodes which join and/or leave the communication range of a given node in the network will certainly change its relationship with its neighbors by detecting new link breakages and/or link additions. This can produce a large number of updates in the routing table of each node in MANET. Furthermore this topology change makes an overhead traffic in the process of route maintenance guaranteed by the implemented routing protocols in MANETs. So, the performance of a MANET is closely related to the capability of the routing protocol to adapt itself to topology changes and the link status [3, 4]. There are three main categories of MANET routing protocols: Proactive (table-driven), Reactive (on-demand) and Hybrid. Proactive protocols build their routing tables continuously by broadcasting periodic routing updates through the network; reactive protocols build their routing tables on demand and have no prior knowledge of the route they will take to get to a particular node. Hybrid protocols create reactive routing zones which are interconnected by proactive routing links and usually adapt their routing strategy to the amount of mobility in the network.

Simulating routing protocols in a network simulator have several benefits over real world testing. Simulations allow researchers to evaluate the performance of routing protocols in a wide range of scenarios at no cost. For this reason, several mobility models were developed to simulate the pattern movement of nodes that will be followed during the simulation [5].

To evaluate MANET protocols, it is not suitable to use only one mobility model. Various models that span across all different mobility characteristics are needed. When evaluating a single protocol, this protocol is run on various models to see how its performance changes on different models. It is found that the performance of a specific protocol varies if underlying mobility models are different. When evaluating a group of protocols, these protocols are run on a single mobility model to see how these protocols behave with this modelled motion. It has been noticed that the behavior of protocols depends on mobility models used. So, it is very difficult to draw a conclusion on the performance of routing protocols from the obtained results depending on the chosen mobility models. Thus, to make this comparison possible, it's necessary to make simulations in the same *mobility condition*. To formulate this mobility condition we used a quantitative metric of mobility [4] that is independent of any used mobility model. This metric measures mobility of network when each simulation takes end.

In this paper, we present a unified quantitative metric of the mobility related to the change of link status. This mobility metric is calculated at regular discrete time intervals. Moreover, it is characterized by two main properties. First, logical because it has a strong linear relationship with the change of link status for a wide range type of scenarios. Second, lightweight because it is light in calculation and recalculation: it is easy to calculate it at regular intervals (during the exchange of HELLO messages) with less consumption in memory and CPU resources, which is suitable for real MANET.

The paper is organized as follows. In Sect. 2, related works are given and discussed. Section 3 presents an overview on mobility models and our proposed metric of mobility. Section 4, gives parameter simulations, main performance metrics and the discussion of the obtained results. The last section concludes the paper and gives an idea about our future works.

2   Related Works

Different mobility models lead to different mobility patterns. But models themselves do not give clear images how mobility patterns are different with each others. We need some mobility metrics to describe these mobility patterns. Efforts to find appropriate mobility metrics have begun only recently. We classify mobility metrics in two categories: direct mobility metrics and derived mobility metrics. The direct mobility metrics, like host speed or relative speed, are measurements with a clear physical meaning. The derived mobility metrics, like graph connectivity, are measurements derived from physical observations through mathematical modelling. The direct mobility metrics measure host motion directly, e.g., average host speed or minimum/maximum speed. For RandomWayPoint Model, pause time is also used to reflect node mobility [3, 6], namely, the longer the pause time is, the smaller the mobility. Other metrics belonging to this category include average relative speed [7], average degree of spatial dependence and temporal dependence [8]. Average relative speed [7] is defined based on relative speed of all pairs of hosts in the network. Attempts have been made in [8] to characterize the temporal dependence of the movement of an individual host and the spatial dependence between different hosts. The temporal dependence indicates how an individual host changes its velocity over time, or say, whether its current velocity is dependent on the previous velocity. Average degree of temporal dependence is proposed to capture temporal dependence. It is an average over the temporal dependence of all the hosts. For each host, the degree of temporal dependence is defined as the product of relative direction and relative speed (relative to itself) at two different times. The spatial dependence indicates whether a hosts movement is correlated with other hosts. The average degree of spatial dependence is the average of degree of spatial dependence over all host pairs. The direct mobility metrics has been used to measure different mobility models. For example, average degree of spatial dependence differentiates different mobility models successfully [8]; average relative speed varies almost linearly with link change rate (see next subsection) under Random WayPoint Model. However, some metrics can not accurately capture different characteristics of the models. For example, average degree of temporal dependence fails to differentiate different mobility models [9]. Average or minimum/maximum speed has been used widely. Although it indicates the degree of mobility; it fails to reflect relative motions between hosts. The metric "pause time" is model dependent: it can only be used in the Random Way-Point Model. More important, direct mobility metrics often do not directly reflect topology changes, while the latter is believed to be more influential to network performance. Take the RandomWalk Model for example; high mobility speed doesn't necessarily generate large geographic movement [10] to cause dramatic topology changes. Mobility models impact the connectivity graph which in turn influences the protocol performance. It is thus helpful to study metrics that capture the properties of connectivity graph. The category of derived mobility metrics include metrics derived from graph theoretic models as well as other mathematical models. Metrics derived from graph-theoretic models include link change rate [10, 11], link duration [8, 12] and path duration [8]. The mobility measure metric proposed in [4] is derived from probabilistic models. Papers [10, 11] proposed link change rate as an indicator of topology change. If a link between two hosts is established/severed due to host movement we consider the state of the link between them up/down. Link change rate is the total number of link up/downs in unit time. Average link duration [8, 12] is defined as the average of link durations over the host pairs that are within each others transmission range. The link duration is the time interval during which two hosts are within each others transmission range. Average path duration [8] averages the durations of all the paths linking every source destination pairs. Path duration is the time interval during which all links on a path (from a source to a destination) exist. The average path duration is related to the path length (hop count) h, average relative speed V and transmission range R. Mobility measure [4] is derived from the average relative speed. It is based on the observation that relative speeds do not make much sense for two nodes that are far away, but make much sense for two nodes that are near the transmission range of each other. A relation of remoteness between two nodes is defined as a function of the distance between two nodes; it increases from 0 to 1 monotonically. The derivative of remoteness is 0 at distance 0, increases as the distance increases, reaches its maxima at the communication boundary; then decreases as distance increases further, and approaches 0 as the distance approaches infinite. The mobility measure is defined as the average of the derivative of remoteness over all node pairs. Evaluations have been performed to investigate how the derived mobility metrics are related to direct mobility metrics, how well the derived metrics can differentiate different mobility models, and how well the metrics can quantify routing performance. Results from [10] show that the link change rate increases as average host speed increases. But results also show that, for different mobility models, differences in link change rate are small, which means link change rate can not differentiate different mobility models effectively. Moreover, [12] pointed out that the drawback of the link change rate is that it only counts the number of link changes without taking into account the duration of a link which heavily influences protocol performance. To this extent, [12] argues that average link duration is a good metric that not only quantifies host movements but also indicates protocol performance accurately. Under Random WayPoint Model, when average link duration increases, throughput increases and end-to-end delay and protocol overhead decreases consistently. For average path duration, it is found that at a high speed, path duration always shows exponential distribution no matter what mobility model is in use; it is also found that there exists a linear relationship between the reciprocal of the average path duration and the routing protocol performance in terms of throughput and routing overhead [8]. For the mobility measure defined on remoteness, simulations show that it has a consistent linear relationship with the link change rate for various mobility models [4].

3 Used Mobility Metric

3.1 Mobility Models Overview

Mobility models are important because they determine the behavior of mobile nodes (MN) on stage [5, 13]. They can be classified into two types: those based on traces (logs of actual movements) and the synthetic (emulate reality by mathematical equations). Some authors classify mobility models into three groups [13]: models based on strokes (work with real mobility), models based on topology restrictions (real scenario simulations) and statistical models (study from randomness). Ad hoc networks do not work yet on models based on traces on the network characteristics. However, it is expected that study will expand in future on the application of these models [5, 13]. Therefore, models of synthetic mobility are used together with simulated scenarios. In order to prove this form of controlled mobility, certain parameters are used, which allow to obtain quantifiable date and thus to transform them into useful informaties. The synthetic models are classified according to their relationship with the representation of human mobility: synthetic mobility models unrealistic, for example: random models [13–15] (Random Walk Mobility Model, Random Waypoint Mobility Model), temporal dependency models [14, 16] (Boundless Simulation Area Mobility Model, Gauss-Markov Mobility Model, Smooth Random Mobility Model) and realistic synthetic mobility models such as: spatial dependence models [16, 17] (Reference Point Group Mobility, Column Mobility Model, Pursue Mobility Model, Nomadic Mobility Model) Geographic Restriction Models [17, 18] (Pathway Models, Obstacle Models, Human Obstacle Mobility Model).

The MNs movement in mobility model activity can be divided into analytical and simulation models. Analytical models can provide performance parameters and Simulation models can derive valuable solutions for more difficult situations. To summarize, categorical mobility model includes:

1. Brownian Model
2. Random Waypoint Model
3. Random Walk Model
4. Random Direction Model
5. Random Gauss-Markov Model
6. Markovian Model
7. Incremental Model,
8. Mobility Vector Model
9. Reference Point Group Model (RPGM)

10. Pursue Model
11. Nomadic Community Model
12. Column Model
13. Fluid Flow Model
14. Exponential Correlated Random Model
15. Map Based Model
16. Manhattan Mobility Model
17. Mission Critical Mobility Model
18. Obstacle Mobility Model
19. Smooth Random Mobility Model
20. Post Disaster Mobility Model

For a given simulation, the chose of Mobility models is based on their different classes of motion as random based and group based movements.

3.2   Used Mobility Metric

In this section, we define our unified metric of mobility that is independent of any used mobility model, and related to the change of link status in the network. This mobility metric is calculated at regular discrete time intervals. Moreover, it is characterized by two main properties. First, logical because it has a strong linear relationship with the change of link status for a wide range type of scenarios. Second, light-weight because it is light in calculation: it is directly based on the data structure of neighbors used by each node in the ad-hoc network, and dont require any supplement process.

Based on the number of nodes leaving and/or joining the communication range of a given node in the MANET, we define two mobility measures in mobile ad-hoc networks. The evaluation of these two metrics can be made at regular time intervals as follows:

$$M_i^\lambda(t) = \lambda \frac{Nodes\,Out(t)}{Nodes(t - \Delta t)} + (1 - \lambda)\frac{Nodes\,In(t)}{Nodes(t)} \tag{1}$$

After the estimation of the node mobility, we define the *wireless network mobility* at regular time intervals as follows:

$$Mob_\lambda(t) = \frac{1}{N}\sum_{i=0}^{N-1} M_i^\lambda(t) \tag{2}$$

where N is the number of nodes in the network. In addition, for a network in steady state, we can use the time average of the mobility measure as follows:

$$M_\lambda = \frac{\Delta t}{T}\sum_k Mob_\lambda(t) \tag{3}$$

where $k \in \{\Delta t, 2 * \Delta t, ..., [T/\Delta t] * \Delta t\}$. The [.] is the integer part and T is the total simulation time.

After presentation of the mobility properties of the proposed four mobility models related to the four mobility spaces including spatial and temporal dependence, relative speed, and geographic restrictions, we then focus on answering the following questions:

1. *Why don't, we get the same performance results through different mobility models of a MANET protocol simulated in the same network parameters?*
2. *What conditions must be considered to compare performance MANET protocols in different mobility models?*

4 Simulation and Results

To answer these questions, we have chosen to evaluate performance of the Optimized Link State Routing (OLSR) protocol [19], one of the proactive MANET routing protocols. Precisely, we have used the Network Simulator NS-2 [20] and the compatible OLSR implementation [21] to measure the three main metrics of performance cited in previous chapter.

4.1 Simulation Environment

This OLSR evaluation was made for two mobility models (RWP and RGM models) in the worst case by supposing that the maximum speed of nodes is equal to Vmax = 40m/s and the pause time is equal to 0 in case of Random Waypoint model. Moreover, simulations are considered in the same MANET environment as illustrated in the Table 1.

Table 1 Simulation parameters used in NS2 for OLSR performance evaluation

Simulation settings	
Nodes number	50 nodes
Topology area	1000 m × 1000 m
Transmission range	100 m
Traffic type	Canstat Bit Rate(CBR)
Connection rate	4 pkts/s
Packet size	512 bytes
Connections number	10
Simulation time	500 s

4.2 Performance Metrics

We have considered the more important metrics for analyzing and evaluating performance of MANET routing protocols. These considered metrics are:

- *Normalized Routing Overhead (NRL)*: It represents the ratio of the control packets number propagated by every node in the network to the data packets number
  received by the destination nodes. This metric reflects the efficiency of the implemented routing protocols in the network.
- *Packet Delivery Fraction (PDF)*: This is the total number of delivered data packets
  divided by the total number of data packets transmitted by all nodes. This performance metric gives us an idea of how well the protocol is performing in terms of
  packet delivery by using different traffic models.
- *Average End-to-End delay (Avg-End-to-End)*: This is the average time delay for
  data packets from the source node to the destination node. This metric is calculated
  by subtracting "time at which first packet was transmitted by source" from "time
  at which first data packet arrived to destination". This includes all possible delays
  caused by buffering during route discovery latency, queuing at the interface queue,
  retransmission delays at the MAC layer, propagation and transfer times.

4.3 Results and Discussion

The Fig. 1 shows that with the same MANET parameters, the RWP mobility model
witnesses less mobility compared to the RGM mobility model.

Fig. 1 OLSR protocol
performance versus Mobility
metric M^λ

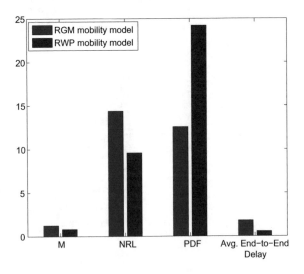

We find out that the three performance metrics are very good in the case of the RWP model, which guarantees less mobility: less messages of control ($NRL(RGM) > NRL(RWP)$), better delivery of data packets ($PDF(RGM) < PDF(RWP)$), and better average end to end delay ($Avg-End-to-End(RGM) > Avg-End-to-End(RWP)$). Moreover, we observe that the performance of the OLSR protocol is sensitive to the mobility in the MANET. Indeed, according to Fig. 1, we notice a light difference between the two mobility measures, but for the three performance metrics, a very significant difference is found:

- $NRL(RGM) \simeq 3/2\ NRL(RWP)$
- $PDF(RGM) \simeq 1/2\ PDF(RWP)$
- $Avg-End-to-End(RGM) \simeq 3\ Avg-End-to-End(RWP)$

Hence, so as to have the same performances of a protocol it is necessary to make simulations in the same *mobility condition*, i.e to chose scenarios with the same mobility measure before beginning simulations.

5 Conclusion

This paper presents a new metric of mobility to evaluate MANETs performance. This mobility metric is unified and quantify the network mobility independently of the nature of the used mobility model. To show how the mobility can impact a MANET performance, we have run simulations under different MANET configurations. After that, we have used this mobility metric to evaluate and compare the MANET performance which uses OLSR protocol as a routing protocol. The considered mobility models in this study are RWP (based random) and RGM (with temporal dependence) models.

References

1. Adam, G., et al.: Performance Evaluation of Routing Protocols for multimedia transmission over Mobile Ad hoc Networks, IFIP WMNC (2011)
2. Jubin, J., Tornow, J.D.: The DARPA packet radio network protocols. Proc. IEEE **75**(1), 21–22 (1987)
3. Jayakumar, G., Gopinath, G.: Performance comparison of two on-demand routing protocols for ad-hoc networks based on random way point mobility model. Am. J. Appl. Sci. **5**(6), 659-664 (2008). ISSN 1546-9239
4. He, L., Yin, W.: A measure of mobility for evaluating mobile ad hoc network performance. In: International Conference on Microwave and Millimeter Wave Technology (2008)
5. Ribeiro, A., Sofia R.: A survey on mobility models for wireless networks, SITI Technical Report SITI-TR-11-01 (2011)
6. Broch, J., Maltz, D.A., Johnson, D.B., Hu, Y.C., Jetcheva, J.: A performance comparison of multi-hop wireless ad hoc network routing protocols. In: Proceedings of the Fourth Annual

ACM/IEEE International Conference on Mobile Computing and Networking (Mobicom 98) (1998)

7. Johansson, P., Larsson, T., Hedman, N., Mielczarek, B., Degrmark, M.: Scenario-based performance analysis of routing protocols for mobile ad-hoc networks. In: Proceedings of the 5th annual ACM/IEEE International Conference on Mobile Computing and Networking, pp. 195–206, Seattle, Washington, USA, Aug 15–19 (1999)

8. Trivio-Cabrera, A., Garca-de-la-Nava, J., Casilari, E., Gonzlez-Caete, F.J.: An analytical model to estimate path duration in MANETs, MSWiM'06, Torremolinos, Malaga, Spain Oct 26 (2006)

9. Bai, F., Sadagopan, N., Helmy, A.: Important: a framework to systematically analyze the impact of mobility on performance of routing protocols for ad hoc networks. In: Proceedings of IEEE Infocom (2003)

10. Hong, X., Kwon, T., Gerla, M., Gu, D., Pei, G.: A mobility framework for ad hoc wireless networks. In: Proceedings of ACM Second International Conference on Mobile Data Management (MDM 2001) (2001)

11. Zhou, B., Xu, K., Gerla, M.: Group and swarm mobility models for ad hoc network scenarios using virtual tracks. In: Proceedings of MILCOM'04, pp. 289–294 (2004)

12. Yawut, C., Paillassa, B., Dhaou, R.: On metrics for mobility oriented self adaptive protocols. In: Proceedings of the Third International Conference on Wireless and Mobile Communications (ICWMC 2007) (2007)

13. Fehnker, A., Hfner, P., Kamali, M., Mehta, V.: Topology-based mobility models for wireless networks. In: 10th International Conference Quantitative Evaluation of Systems QEST 2013, Buenos Aires, Argentina, Aug 27–30 (2013)

14. Divecha, B., Abraham, A., Grosan, C., and Sanyal, S.: Impact of node mobility on manet routing protocols models. J. Digit. Inf. Manag. (2007)

15. Pal, N., Dhir, R.: Analyze the impact of mobility on performance of routing protocols in MANET using OPNET modeller. Int. J. Adv. Res. Comput. Sci. Softw. Eng. 3(6), (2013)

16. Hong, X., Gerla, M., Pei, G., Chiang, C.-C.: A group mobility model for ad hoc wireless networks. In: Proceedings of the 2nd ACM international workshop on Modeling, Analysis and Simulation of Wireless and Mobile Systems, MSWiM 99, p. 5360, ACM, New York, USA (1999)

17. Chenchen, Y., Xiaohong, L., Dafang, Z.: An obstacle avoidance mobility model. In: IEEE International Conference on Intelligent Computing and Intelligent Systems (ICIS), vol. 3, pp. 130–134 (2010)

18. Aschenbruck, N., Gerhards-Padilla, E., Martini, P.: A survey on mobility models for performance analysis in tactical mobile networks. J. Telecommun. Inf. Technol. (2008)

19. Clausen, T., Jacquet, P., Adjih, C., Laouiti, A., Minet, P., Muhlethaler, P., Qayyum, A., Viennot, L.: Optimized link state routing protocol (olsr) Rfc 3626 (2003)

20. NS2, official http://www.isi.edu/nsnam/ns/ (2018)

21. ROS, F.J.: Um-olsr version 8.8.0, university of murcia, spain. http://masimum.dif.um.es/?software:um-olsr (2018)

Internet of Vehicles Over Named Data Networking: Current Status and Future Challenges

Chaker Abdelaziz Kerrche, Farhan Ahmad, Mohamed Elhoseny, Asma Adnane, Zeeshan Ahmad and Boubakr Nour

Abstract Vehicular ad hoc networks (VANETs) have emerged as a new breed of Self-Organized Networks (SONs). In addition to the Vehicle-to-Vehicle and Vehicle-to-RoadSide Units, it assumes that vehicular nodes are connected to other smart objects equipped with a powerful multi-sensor platform, communication technologies, computation units, and IP-based connectivity to the Internet, thereby creating a network called Internet of Vehicles. IoV applications are also based on cooperation-aware communication where information is exchanged among communicating nodes. This phenomenon advocates for content-centric communication paradigm for IoV applications where the information is of the essence than the source and/or location of information. As a result, owing to the Future Internet Architecture (FIA), recently Named Data Networking (NDN) have been considered to be the architectural backbones for SONs and its breeds. In this work, we provide details of IoV over NDN. Specifically, we focused on the architectural details and requirements of NDN-enabled IoV. We also discuss the current status of the realization of NDN-enabled IoV. Finally, we point out some of the research challenges in this domain.

C. A. Kerrche (✉)
University of Ghardaia, Bounoura, Algeria
e-mail: kr.abdelaziz@gmail.com

F. Ahmad
University of Derby, Derby, UK

M. Elhoseny
Mansoura University, Mansoura, Egypt

A. Adnane
Loughborough University, Loughborough, UK

Z. Ahmad
King Khalid University, Abha, Kingdom of Saudi Arabia

B. Nour
Beijing Institute of Technology, Beijing, China

© Springer Nature Switzerland AG 2020
M. Elhoseny and A. E. Hassanien (eds.), *Emerging Technologies for Connected Internet of Vehicles and Intelligent Transportation System Networks*, Studies in Systems, Decision and Control 242, https://doi.org/10.1007/978-3-030-22773-9_7

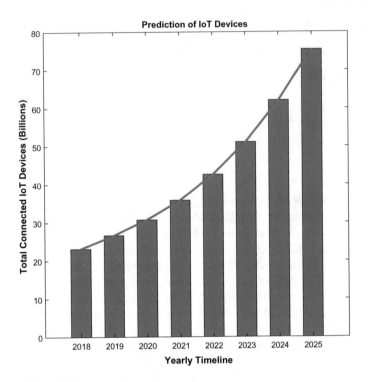

Fig. 1 Prediction of IoT devices [2]

1 Introduction

Internet-of-things (IoT) has recently emerged as a novel and innovative technology, which aims to provide Internet connectivity to every device having computation, communication and storage ability. This phenomenon of connecting smart devices with Internet results in numerous applications in different domains including smart health, smart cities, smart industries, smart homes, to name a few [1]. As a result, it attracts various industries, manufacturers and venders from across the globe due to to the massive range of application IoT supports. According to recent statistic, there are currently 23.14 billion IoT devices worldwide and it is likely to be increased threefold in 2025 as depicted in Fig. 1 [2]. Further, GSMA predicted that IoT is on track to generate about 1.1 trillion dollars worldwide by 2025, thus, providing a massive opportunity for different stakeholders including vendors, enterprises and manufacturers to invest in such innovative technology [3].

In order to provide secure and safe transportation, IoT has evolved into a novel technology, known as IoV where smart vehicles equipped with computing, communicating and storing abilities communicates with each other and surrounding infrastructure via different modes of communication including vehicle-to-vehicle (V2V), vehicle-to-infrastructure (V2I), vehicle-to-pedestrian (V2P), and infrastructure-to-infrastructure (I2I) [4]. This enables the smart vehicles to exchange critical information including sudden lane changes, steep-curves, black ice warnings, and traffic accidents avoidance, thus assisting the drivers in extreme and difficult situations. According to SBD automotive, there will more than 81 million vehicles around the globe by 2025, out of which 68% vehicles will offer some sort of connectivity [5]. IoV has a massive potential to provide secure and safer transportation as it enables the vehicles to take correct decision in a timely manner. Further, applications of IoV include better fuel consumption, providing infotainment to vehicular users and optimizing journey times in an efficient manner [6, 7]. In a nutshell, IoV is an extremely significant technology, which can increase traffic efficiency and provides safety to users on the road.

IoV can succeed only, if the transmission of information from one vehicle can be delivered successfully to other vehicles in a secure and trusted environment. This information delivery mechanism in current IoV follows a point-to-point (P2P) communication pattern where messages are shared with vehicles having IP addresses. However, the applications in IoV are mostly point-to-multipoint (P2M) in nature, where the messages generated from one vehicle is broadcasted and shared with vehicles in its neighbourhood. Moreover, IP-based communication violates the traditional IoV security attributes as they are more prone to user attacks and data losses.

Named Data Networking (NDN) [8] is recently proposed as an alternative Internet architecture which can address the issues of IP-based Internet architecture [9]. NDN follows the principle that content is more vital than the content provider, i.e., it shifts the focus of Internet architecture from the concept of "where" to "what". Due to this nature, NDN fully complies with the nature of IoV, where transmitted data is of extreme importance than the location of the data generator. NDN-based IoV enables the vehicles to access the critical information within the network. Based on this critical information, vehicles can take correct decisions in time to ensure transportation security and safety.

This chapter is dedicated to provide details of IoV over NDN. Specifically, we focused on the architectural details and requirements of NDN-enabled IoV. Further, we also discussed the current status of the realization of NDN-enabled IoV. Moreover, we also discussed some of the research challenges in this domain.

The rest of this chapter is organized as follows: Sect. 2 explains how future internet architectures emerged due to the emergence of new applications and needs for internet users. Afterwards, we discuss the existing vehicular named data networking solutions in Sect. 3. Section 4 will be dedicated to the open challenges and issues in VNDN domain and finally, Sect. 5 concludes the chapter.

2   From Host-Centric to Content-Oriented Communication

Today's Internet architecture is an entity-oriented design based on TCP/IP protocol, which needs IP addresses of the nodes to establish communication between different nodes. In order to obtain any content over internet, the content provider address is needed. But due to technology advancement over the last decade, internet has become more content-oriented by providing a platform to share contents regardless of who is providing it. Most of the today's applications are content-oriented and are based on principle of what data is needed rather than where it is available [10]. A middleware is used to map this application model with current internet model. So there is a need for a new internet architecture which is content-centric rather than host centric. It should remove all the middlewares, associated configurations and addresses current internet challenges like scalability, security and privacy etc.

With these new challenges come new solutions. One of the proposed future Internet architecture projects is *Information Centric Networking* (ICN) [11]. Several architectures were proposed under the ICN project. *Named Data Networking* is the promising instantiation of the ICN.

In the framework of future internet technologies, Named Data Network (NDN) has been recently proposed [12]. It is is a candidate architecture for Future Internet Architecture (FIA) to replace traditional TCP/IP Internet architecture [12, 13]. As mentioned before, NDN is an extension of an earlier project in the domain of FIA named Content-Centric Networking (CCN), which was first introduced in 2006. NDN architecture relies on the data-centric paradigm where data is the pinnacle of the communication architecture rather than the hosts. NDN uses data names instead of IP addresses for communication, which are hierarchically structured and meaningful, e.g., this paper may have the name: `"Springer/Chapters/ndn/Internet of-Vehicles-over-Named-Networking-Current-Status-and-Future-Challenges"`

As the name suggests, NDN architecture is based on the *named data* (content name) rather than *named devices* (host name). NDN uses two types of packets for communication.

1. *Interest* Packet: Content Consumer initiates interest packets to request some content identified by a unique content name. This request will be broadcast by consumer node.
2. *Data* Packet: The consumer's request is responded by the data packet that carries the requested content and is a unicasted from the content provider towards content consumer.

Typically in NDN architecture, each node maintains the following three data structures.

1. Content Store (CS): It acts as a node's cache. It will store the data content available at the node based on caching policy of node. The main idea is to serve future content requests instead of forwarding the requests to original content provider. It in turns can help to make communication more bandwidth efficient by speeding up the content retrieval process.
2. Pending Interest Table (PIT): It keeps track of all the current interest requests which are not satisfied. Each PIT entry stores interest request details that includes a content name, the interface through which interest request is received, and timers for managing PIT entry.

3. Forwarding Information Base (FIB): It helps in forwarding the current unsatisfied interest requests to the next hop node, upstream towards the content provider. The next hop node is chosen by the current node based on the longest name prefix match. Each FIB entry contains the name prefixes and the interfaces through which interest request is forwarded.

As discussed earlier, NDN uses interest and data packets for communication. This communication process is performed in two stages. First stage includes forwarding the requested interest packet towards the content provider. Whereas, second stage function is to deliver back the requested data packet to the requesting node, as discussed below.

1. A content consumer in need of a certain content, will initiate and generate an interest packet, to request the content based on unique content name [14]. This interest packet will be then forwarded by the node in a hop-by-hop fashion upstream towards the content provider. The intermediate hop nodes will act as relay to forward the interest packet to the content provider or next-hop node, if they do not have the requested content. Any node when receives the interest packet will perform the following steps.

    (a) First of all, it will check the requested content name with the content it already cached in its local CS. If the requested content matches the content in local CS, it will send back a copy of the content through the interface it received the request and then discards the interest packet.
    (b) In case the content is not available in CS, node will search the PIT to find out if any other node also requested the same content and its request has not been satisfied yet. The incoming unsatisfied request will be looked up in PIT using content name. If PIT look up match is found, the incoming interest packet interface will be appended to already existing PIT entry. This aggregation of multiple interest requests for same content will help to the node to deliver back the data packet to all requested nodes, once the current node receives the data packet. The interest packet will be discarded after aggregation.
    (c) In case, PIT look up match is not found, the node will look up FIB to find out suitable next hop node based on the longest name prefix. Once it finds out the match, it will forward the interest packet to the selected next hop node and record a new entry in PIT and FIB with the interest name and incoming/outgoing interfaces.

2. Once the node receives any data packet, it will first look up data packet name in PIT to find out if it is a request data packet.

    (a) If name data packet name does not match with the PIT entry, the node will discard this packet assuming it is not been requested.
    (b) If node finds the name match with the PIT entry, it will forwards the data packet to all the interfaces registered with that particular PIT entry. Once the interest packet request is fulfilled, the node will then remove its entry from the PIT. Also, the node will decide to cache the copy of the data packet based on the caching policy it implements.

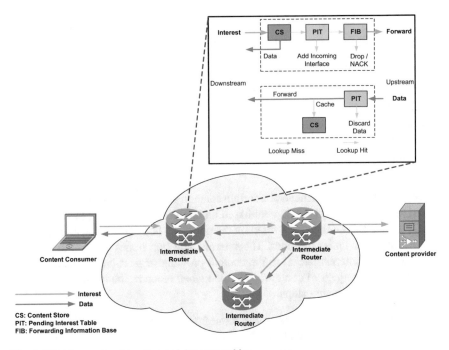

Fig. 2 Working mechanism of named data networking

Figure 2 highlights and summarizes the working mechanism of NDN in both upstream and downstream direction. The next section focuses on the evolution and realization of NDN for one of the significant IoT application, i.e., VANET.

3 Vehicular NDN Solutions

In the era of intelligent transportation systems, many new applications and new types of communications have been made possible, these applications rely on an efficient, reliable and secure content/data distribution and delivery platforms via vehicles and RSU.

The concept of NDN seems to fit perfectly the philosophy of many VANET applications [15]. Indeed, nodes (vehicles) are more likely to be looking for information (data-centric model) about an area rather than having a peer to peer connection with another vehicle (node-centric model) for several applications such as: parking, weather forecast e-advertisement, traffic jam updates and journey planning [16]. Several research have been undertaken to study the feasibility of applying content-centric approach for vehicular networks [17–19] and few NDN-based architectures have been suggested for VANET [20–25].

As mentioned previously, NDN uses data names for data delivery. Name spaces for all applications should be defined and shared with a clear structure *a priori*. In fact, structured and hierarchical names must be meaningful to high level applications and uses so that an interest can be easily processed. Consequently, nodes can fetch data from any neighbouring nodes. However, this becomes challenging in highly mobile network such as VANETs when the data source is not necessarily available in the neighbourhood, vehicles need to decide about forwarding interests for the requested content. Moreover, It is important to mention that NDN delivery concept provides several functions that are difficult in IP based delivery, such as in-network caching, multicast delivery, multi-path forwarding, and data provenance.

For an efficient VANET NDN-based architecture, all the NDN elements should be considered carefully in the design: NDN naming, NDN routing/forwarding, NDN caching, NDN security/privacy and mobility; which will be discussed in the following subsections. Indeed, various research solutions have proposed different NDN architectures for VANET (e.g. Fig. 3 [15]) where vehicles can be consumers, providers, and/or data forwarders. At the same time, RSUs and core network elements can provide NDN forwarding and caching capabilities.

Figure 3 illustrates a generic NDN architecture for VANET applications. The architecture consists of different layers such as physical, strategy, NDN, security, and application layers. The NDN layer is an intermediate layer providing the core NDN aspects (naming, caching, mobility, and routing). Similar to other suggested VNDN architectures, all VANET communication interfaces (DSRC, WiFi, and 3G/4G) are used in NDN to ensure more availability of the data across the network. Furthermore, it is important to take into account the nature of the application handled (*Application layer*), which has shaped somehow certain suggested architectures in the literature. Indeed, an VNDN architecture for time sensitive applications (e.g. news, road traffic, emergency) will not focus on same QoS requirements as other applications. Thus, it is important to have a VNDN architecture that takes into account the different categories of VANET applications and ensure the best outcomes for the requirements of those applications.

3.1 VNDN Forwarding Strategies & Mobility

Since in NDN the target is the data rather than it's owner, identity-based routing and forwarding strategies cannot perform. Hence, new solution were proposed.

Authors of [26] introduced two variants of a new cache decision and replacement policy for NDN that take into account content popularity. Furthermore, they have implemented the Vehicular named data network model in two networks scenario: V2I and V2R -V2V.

NAVIGO is another scheme proposed in [27], it aims to explore the area surrounding the node looking for producer, mules or RSU and then, as soon as the first Data packet comes back, forwarding future Interests for the same prefix towards the

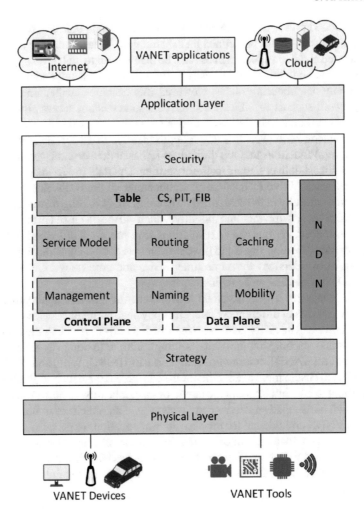

Fig. 3 A reference NDN architecture in VANET [15]

geo-area where the data is coming from. So it binds data names to the producers' geographic areas.

Ahmed et al. [28] proposed a solution in which, each consumer/forwarder can select only one vehicle among the immediate neighboring vehicles for interest forwarding. Each vehicle maintains a local data structure which contains the list of satisfied interests information by that particular vehicle. All neighbors store this information in their Neighbors Satisfied List (NSL), which helps to select the potential interest forwarder.

Authors of [29] applied named data to the on-going vehicle networks and described a prototype implementation of the V-NDN. A database naming The traffic information dissemination application was developed in [30]. It depicts that names

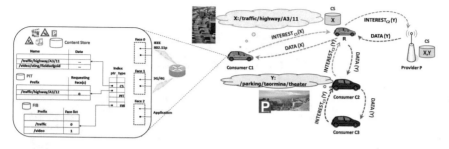

Fig. 4 Example of VNDN forwarding strategy [34]

can greatly facilitate the dissemination process. But use the broadcast to propagate packets can potentially result in poor performance. In addition, kuai and. Al. rated IntPkt published in [31] and indicated that there is an increase in the loss rate in high density scenarios in V-NDN. The work in [32] emphasizes that most vehicle applications focus on obtaining POI information and proposes an approach for mapping two-dimensional geographical areas. Instead of a blind flood, NAIF selects co-operative nodes to transfer the IntPkt fractions. Read and. Al. [33] presented a social link content recovery algorithm based on the use of the K-medium classification algorithm to structure hierarchical architecture between nodes, but it takes a process to build social links. Figure 4 shows an example of VNDN forwarding strategy.

3.2 VNDN Cache Management

Caching means to save data at some location(s) and then use next time if needed [35]. It can be defined as having information, data and object temporarily saved in a location for predictive usage on frequent or closely related interval [36]. Same as in wired NDN, Caching in VNDN can be either coordinated using a mobile cluster head or a road side unit, or a non-coordinated fully distributed caching which is mostly based on the principle of 'Leave Copy Everywhere' [37–39]. Generally, the factors considered in the caching decision are the frequency (popularity) of the requested data, its recency, its cost of retrieval, and its size. Figure 5 shows a classification of the existing cache management solutions.

Besides the caching related question which are: What part of the content is to be cached? When is the most appropriate time for caching? and How would the object be cached? Various additional problems arise with the in-VANET caching due to the environment high mobility and unpredicted topology.

Work in [24] introduces a proactive caching scheme based on mobility prediction in vehicular networks. The proposed scheme uses Long Short Time Memory module in order to predict the right placement to fetch the content in the right time. However, authors target only V2I communication without considering other V2V scenarios.

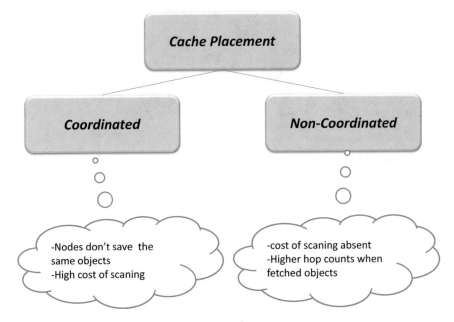

Fig. 5 Cache management in VNDN

A mobility-aware caching strategy for VNDN was proposed in [40]. Based on formed clusters of vehicles moving with similar mobility patterns this cooperative caching approach mainly aim at mitigating the impact of vehicle mobility, as the link between nodes with a similar pattern is relatively stable and reliable.

Work in [38] propose a multi-metric content store management mechanism through cache replacement. The considered three main metrics methods namely: freshness of the content, the frequency of retrieval (popularity), and the distance between the location where the content was received/saved in CS and the current location of the caching node. Simulation results depicts that this proposal is the most efficient VNDN caching strategy in terms of Hit Ratio.

In another work [41], a Distributed Probabilistic Caching (DPC) strategy in VNDN, each node makes the caching decisions from interest entries, Degree and Betweenness Centrality, and relative movement of the receiver and the sender. In [42], authors discussed the cache policy in an ICN (V2V) scenario. They evaluate the community similarity and privacy rating of vehicles, selects the caching vehicle based on content popularity, and put forwards the Popularity Prediction-based Cooperative Cache Replacement (PPCCR) mechanism. For a Vehicle-to-Infrastructure scenario authors of [43] presented an Integer Linear Programming (ILP) formulation of the problem of optimally distributing contents in the network's nodes. In [44], authors propose a ICN-based COoperative Caching solution (ICoC) for a V2V scenario to improve the quality of experience (QoE) of multimedia streaming services.

In addition to strategies to identify where to cache, management strategies as part of the caching nodes, describing how to deal with content to cache at the node, also have a significant impact on the performance of the network. Such strategies are similar to cache replacement strategies [45], Least Recently Used (LRU) the last used content is discarded [46], Least Frequently Used (LFU) cache replacement policy because it gets rid of the contents that are less frequently used first. In [47], RaNDom (RND) study the performance of the Random (RND) replacement policy where the item to be evicted is randomely chosen and replaced with the new item.

Finally, authors of [48] came up with a scheme of cache replacement that depends on the popularity of content which it is Cache Content Popularity (CCP).

3.3 VNDN Security

Work in [49] reviews the existing VANET attacks and classifies them based on NDN point of view. The authors define different issues and challenges in order to address them in the future by the research community. Similarly, work in [50] discusses the security and attacks in vehicular cyber-physical systems by highlighting various issues and challenges. The authors propose an NDN-based cyber-resilient architecture that contains NDN forwarding daemon, threat aversion, detection, and resilience provision.

Work in [51] proposes a mitigation technique to avoid cache poisoning attack in NDN-based network. A trust model has been designed that take both credibility, and feedback into consideration to identify which content is invalid. While [52] proposes a reputation-based Blockchain scheme in order to secure the data caching in vehicular networks. The proposed scheme is based on using a reputation value for the served cache stores that is increasing only if the cache store served a trust and valid contents. However, the use of Blockchain is still a challenging for VANET applications that require a short response delay and a real-time data retrieval.

Taking the privacy factor into design, work in [53] introduces a trust model for autonomous vehicular applications based on NDN. The main contribution of this solution is to prevent the false data and vehicle tracking. A four-level hierarchical trust model has been presented that includes *autonomous-Vehicle* organizations, *manufacturers*, *Vehicles*, and *data*. The authors also combine a hierarchical naming scheme with the model to detect false information.

Besides, many security issues still need to be addressed. Figure 6 summarizes the security attacks facing VNDN different layers.

4 Challenges and Future Directions

Although the aforementioned advantages and solutions, there are still some open issues and challenges that need to be addressed before NDN being widely used in vehicular networks. In this section, we discuss the major challenges for vehicular named networks and highlight different guidelines and research directions.

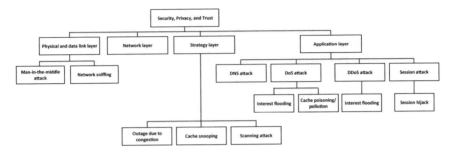

Fig. 6 Security issues facing VNDN different layers

4.1 Content, Data, and Service Naming

One of the most important networking aspects in NDN is the naming, not only because it is the identifier of content but also the pillar element for other functionalities (e.g., forwarding, caching, mobility, security, etc). Names must be short and carry enough information for reliable routing. Furthermore, it must identify devices to provide monitoring and management functions, and provide services and tasks identification to satisfy user's and applications' quality of service requirements.

4.2 Mobility Support

Vehicular networks are considered as highly mobile environments with unpredictable changes in topology and vehicle moves. Mobility support in NDN is a key requirement toward a sable and reliable system. Although NDN provides a simple re-issue mechanism for unsatisfied interests, this may help resolve the issue. Indeed, a scalable naming resolution system to handle the new location of the content producer or an efficient forwarding mechanism to update the pending interest entries are extremely desirable. Moreover, the mobility of cache store must be treated with the same priority as the produce mobility, due to the fact the cache store may be the only content producer in case of original provider unreachability.

4.3 Blockchain-Based Secure VNDN

Recently, Blockchain (BC) is proposed as a decentralized and auditable network where every participating node can add or modify data only after solving a complex puzzle, thus, providing a security-by-design architecture. In BC-based traditional VANET, smart vehicles can only add and amend the block data after solving a Proof-of-Work (PoW) consensus algorithm, thus enabling them to disseminate legitimate content in the network. Though BC can be helpful to address some of the issues of IP-based network, but it will increase the network complexity in terms of scalability. For VNDN, BC can be a revolution as it provides security by design due to its decentralized nature. However, to the best of our knowledge, BC is not utilized within VNDN.

4.4 Social Awareness

The proliferation of handheld devices requires mobile carriers to provide anytime and anywhere connectivity. The mobility patterns of mobile devices strongly depend on the users' movements, which are closely related to their social relationships and behaviors. Consequently, today's mobile networks are becoming increasingly human centric. This leads to the emergence of a new field known as socially-aware networking [54]. Only few works have considered the social relationship aming moving vehicles known as Vehicular Social Networks (VSNs) to improve the different NDN issues including caching, security, and forwarding strategies [55]. Hence, NDN-based VSNs is one of the major open research directions.

4.5 NDN-Based Vehicular Clouds

Vehicles become smarter objects, able to share their treatment, storage and detect resources to support advanced services (for example, merging and processing data from different sensors for standalone driving), acting as a local cloud. However, the implications and outstanding issues of NDN-based systems vehicle clouds are still numerous, as NDN should evolve from a framework that provides content to a system of to orchestrate heterogeneous and complex tasks. This means that the semantics of the NDN packets, the transmission and the cache fabric needs to be rethought.

4.6 Cooperative Applications

Self Organized Networking (SONs) like VANET are generally built to support cooperative applications including safety and infotainment applications. They are generally based on periodically exchanged and event triggered messages. Moving to the data-centric architecture opens various new problems. To date, most of VNDN cooperative applications focus on the caching strategies. However, cooperative safety, online games, and video streaming are among the cooperative applications requiring more investigation in the era of VNDN and IoV.

4.7 Technology Heterogeneity

Heterogeneous technologies are generally used for different purposes: the cellular network is used to carry the NDN Interests over long distances and counteract possible connectivity gaps over the V2V links, while short-range V2V communications are exploited for content delivery. Besides, Software Defined Networking are (SDHVNet) architecture are also recently considered for ensuring a highly agile networking infrastructure to ensure rapid network innovation on-demand. Thus, further investigations are required to find the suitable technology for which VNDN can offer better performance.

4.8 Business Models

The current Internet business model cannot be adapted for NDN paradigm since the latter is based on host-connectivity while the later based on content-oriented communication. It is very important to develop new business models that can deal with long-term NDN business requirements and the new player in the system (i.e., cache store). The desirable model must motivate different entities in the network to allow their data to be transparently cached regardless of who will cache it, and provide local resources such as storage and computation for external usage.

5 Conclusion

Internet of Vehicles (IoV) applications are also based on cooperation-aware communication where information is exchanged among communicating vehicles. This phenomenon advocates for a content-centric communication paradigm for IoV applications where the focus is on information itself rather than the source and/or location of information. As a result, owing to the Future Internet Architecture (FIA), Information Centric Networking (ICN), Content-Centric Networking (CCN), and Named Data Networking (NDN) have recently been considered to be the architectural backbones for SONs and its breeds. Named Data networking (NDN) is an approach to evolve the Internet infrastructure to directly support this new paradigm by introducing uniquely named data as a core Internet principle. Data becomes independent from location, application, storage, and means of transportation, enabling in-network caching and replication. In this chapter, we provided details of IoV over NDN. Specifically, we focused on the architectural details and requirements of NDN-enabled IoV. Further, we also discussed the current status of the realization of NDN-enabled IoV. Moreover, we also discussed some of the research challenges in this domain.

References

1. Mehmood, Y., Ahmad, F., Yaqoob, I., Adnane, A., Imran, M., Guizani, S.: Internet-of-things based smart cities: recent advances and challenges. IEEE Commun. Mag. **55**(9), 16–24 (2017). https://doi.org/10.1109/MCOM.2017.1600514
2. Statista: IoT: Number of connected devices worldwide 2012–2025, available online: https://www.statista.com/statistics/471264/iot-number-of-connected-devices-worldwide/ (Accessed 30th Nov 2018)
3. Bains, K., Giles, M., Rogers, M., Wyrzykowski, R., Kechiche, S.: IoT: The $1 Trillion revenue opportunity, available online: https://www.gsmaintelligence.com/research/2018/05/iot-the-1-trillion-revenue-opportunity/670/ (Accessed 30th Nov 2018)
4. Yang, F., Wang, S., Li, J., Liu, Z., Sun, Q.: An overview of internet of vehicles. China Commun. **11**(10), 1–15 (2014)
5. SBD: Connected car global forecast. Technical report, SBD automotive (2015), available at: https://www.sbdautomotive.com/files/sbd/pdfs/536 (Accessed 29th Nov 2018)
6. Lu, N., Cheng, N., Zhang, N., Shen, X., Mark, J.W.: Connected vehicles: solutions and challenges. IEEE Internet Things J. **1**(4), 289–299 (2014)
7. Ahmad, F., Ahmad, Z., Kerrache, C.A., Kurugollu, F., Adnane, A., Barka, E.: Blockchain in internet-of-things: architecture, applications and research directions. In: International Conference on Computer and Information Sciences (ICCIS) (2019) [In Press]
8. Zhang, L., Afanasyev, A., Burke, J., Jacobson, V., Crowley, P., Papadopoulos, C., Wang, L., Zhang, B., et al.: Named data networking. ACM SIGCOMM Comput. Commun. Rev. **44**(3), 66–73 (2014)
9. Named data networking: NDN project overview, available online: https://named-data.net/project/ (Accessed 28th Nov 2018)
10. Afanasyev, A., Burke, J., Refaei, T., Wang, L., Zhang, B., Zhang, L.: A brief introduction to named data networking. In: MILCOM 2018–2018 IEEE Military Communications Conference (MILCOM), pp. 1–6. IEEE (2018)
11. Xylomenos, G., Ververidis, C.N., Siris, V.A., Fotiou, N., Tsilopoulos, C., Vasilakos, X., Katsaros, K.V., Polyzos, G.C.: A survey of information-centric networking research. IEEE Commun. Surv. Tutor. **16**(2), 1024–1049 (2014)
12. Zhang, L., Estrin, D., Burke, J., Jacobson, V., Thornton, J.D., Smetters, D.K., Zhang, B., Tsudik, G., Massey, D., Papadopoulos, C., et al.: Named data networking (NDN) project. Relatório Técnico NDN-0001, Xerox Palo Alto Research Center-PARC 157, 158 (2010)
13. Jacobson, V., Smetters, D.K., Thornton, J.D., Plass, M.F., Briggs, N.H., Braynard, R.L.: Networking named content. In: Proceedings of the 5th International Conference on Emerging Networking Experiments and Technologies, pp. 1–12. ACM (2009)
14. Nour, B., Sharif, K., Li, F., Moungla, H., Liu, Y.: M2HAV: a standardized ICN naming scheme for wireless devices in internet of things. In: International Conference on Wireless Algorithms, Systems, and Applications (WASA), pp. 289–301. Springer International Publishing (2017)
15. Khelifi, H., Luo, S., Nour, B., Moungla, H., Faheem, Y., Hussain, R., Ksentini, A.: Named data networking in vehicular ad hoc networks: state-of-the-art and challenges. IEEE Commun. Surv. Tutor. (2019)
16. Saxena, D., Raychoudhury, V., Suri, N., Becker, C., Cao, J.: Named data networking: a survey. Comput. Sci. Rev. **19**, number=, pages=15–55, https://doi.org/10.1016/j.cosrev.2016.01.001 (2016)
17. Amadeo, M., Campolo, C., Molinaro, A.: Content centric networking: is that a solution for upcoming vehicular networks? In: Proceedings of the Ninth ACM International Workshop on Vehicular Inter-networking, Systems, and Applications, pp. 99–102 (2012)
18. Arnould, G., Khadraoui, D., Habbas, Z.: A self-organizing content centric network model for hybrid vehicular ad-hoc networks. In: Proceedings of the First ACM International Symposium on Design and Analysis of Intelligent Vehicular Networks and Applications, DIVANet, pp. 15–22 (2011)

19. Grassi, G., Pesavento, D., Pau, G., Vuyyuru, R., Wakikawa, R., Zhang, L.: Vanet via named data networking. In: 2014 IEEE Conference on Computer Communications Workshops (INFOCOM WKSHPS), pp. 410–415 (2014)

20. Amadeo, M., Campolo, C., Molinaro, A.: Crown: content-centric networking in vehicular ad hoc networks. IEEE Commun. Lett. **16**(9), 1380–1383 (2012)

21. Chen, M., Mau, D.O., Zhang, Y., Taleb, T., Leung, V.C.: Vendnet: vehicular named data network. Veh. Commun. **1**(4), 208–213 (2014), http://www.sciencedirect.com/science/article/pii/S221420961400045X

22. Grassi, G., Pesavento, D., Pau, G., Zhang, L., Fdida, S.: Navigo: Interest forwarding by geolocations in vehicular named data networking. In: 2015 IEEE 16th International Symposium on A World of Wireless, Mobile and Multimedia Networks (WoWMoM), pp. 1–10 (2015)

23. Zhang, Y., Afanasyev, A., Burke, J., Zhang, L.: A survey of mobility support in named data networking. In: 2016 IEEE Conference on Computer Communications Workshops (INFOCOM WKSHPS), pp. 83–88 (2016)

24. Khelifi, H., Luo, S., Nour, B., Sellami, A., Moungla, H., Naït-Abdesselam, F.: An optimized proactive caching scheme based on mobility prediction for vehicular networks. In: IEEE Global Communications Conference (GLOBECOM), pp. 1–6. IEEE (2018)

25. Barka, E., Kerrache, C., Hussain, R., Lagraa, N., Lakas, A., Bouk, S.: A trusted lightweight communication strategy for flying named data networking. Sensors **18**(8), 2683 (2018)

26. Chen, M., Mau, D.O., Zhang, Y., Taleb, T., Leung, V.C.: Vendnet: vehicular named data network. Veh. Commun. **1**(4), 208–213 (2014)

27. Grassi, G., Pesavento, D., Pau, G., Zhang, L., Fdida, S.: Navigo: interest forwarding by geolocations in vehicular named data networking. In: 2015 IEEE 16th International Symposium on A World of Wireless, Mobile and Multimedia Networks (WoWMoM), pp. 1–10. IEEE (2015)

28. Ahmed, S.H., Bouk, S.H., Kim, D.: Rufs: robust forwarder selection in vehicular content-centric networks. IEEE Commun. Lett. **19**(9), 1616–1619 (2015)

29. Grassi, G., Pesavento, D., Pau, G., Vuyyuru, R., Wakikawa, R., Zhang, L.: Vanet via named data networking. In: 2014 IEEE Conference on Computer Communications Workshops (INFOCOM WKSHPS), pp. 410–415. IEEE (2014)

30. Wang, L., Afanasyev, A., Kuntz, R., Vuyyuru, R., Wakikawa, R., Zhang, L.: Rapid traffic information dissemination using named data. In: Proceedings of the 1st ACM Workshop on Emerging Name-Oriented Mobile Networking Design-Architecture, Algorithms, and Applications, pp. 7–12. ACM (2012)

31. Kuai, M., Hong, X., Flores, R.R.: Evaluating interest broadcast in vehicular named data networking. In: 2014 Third GENI Research and Educational Experiment Workshop, pp. 77–78. IEEE (2014)

32. Pesavento, D., Grassi, G., Palazzi, C.E., Pau, G.: A naming scheme to represent geographic areas in NDN. In: 2013 IFIP Wireless Days (WD), pp. 1–3. IEEE (2013)

33. Lu, Y., Li, X., Yu, Y.T., Gerla, M.: Information-centric delay-tolerant mobile ad-hoc networks. In: 2014 IEEE Conference on Computer Communications Workshops (INFOCOM WKSHPS), pp. 428–433. IEEE (2014)

34. Amadeo, M., Campolo, C., Molinaro, A.: Information-centric networking for connected vehicles: a survey and future perspectives. IEEE Commun. Mag. **54**(2), 98–104 (2016)

35. Nour, B., Sharif, K., Li, F., Moungla, H., Kamal, A.E., Afifi, H.: NCP: a near ICN cache placement scheme for IoT-based traffic class. In: IEEE Global Communications Conference (GLOBECOM) (2018)

36. Robinson, J.T., Devarakonda, M.V.: Data Cache Management Using Frequency-Based Replacement, vol. 18. ACM (1990)

37. Lim, H., Ni, A., Kim, D., Ko, Y.B., Shannigrahi, S., Papadopoulos, C.: NDN construction for big science: lessons learned from establishing a testbed. IEEE Netw. **32**(6), 124–136 (2018)

38. Ostrovskaya, S., Surnin, O., Hussain, R., Bouk, S.H., Lee, J., Mehran, N., Ahmed, S.H., Benslimane, A.: Towards multi-metric cache replacement policies in vehicular named data networks. In: 2018 IEEE 29th Annual International Symposium on Personal, Indoor and Mobile Radio Communications (PIMRC), pp. 1–7. IEEE (2018)

39. Ruan, Z., Luo, H., Lin, W.: Enhancing named-based caching in NDN. In: International Conference on Cloud Computing and Security, pp. 320–330. Springer (2018)
40. Huang, W., Song, T., Yang, Y., Zhang, Y.: Cluster-based cooperative caching with mobility prediction in vehicular named data networking. IEEE Access (2019)
41. Deng, G., Wang, L., Li, F., Li, R.: Distributed probabilistic caching strategy in vanets through named data networking. In: 2016 IEEE Conference on Computer Communications Workshops (INFOCOM WKSHPS), pp. 314–319. IEEE (2016)
42. Zhao, W., Qin, Y., Gao, D., Foh, C.H., Chao, H.C.: An efficient cache strategy in information centric networking vehicle-to-vehicle scenario. IEEE Access **5**, 12657–12667 (2017)
43. Mauri, G., Gerla, M., Bruno, F., Cesana, M., Verticale, G.: Optimal content prefetching in NDN vehicle-to-infrastructure scenario. IEEE Trans. Veh. Technol. **66**(3), 2513–2525 (2017)
44. Quan, W., Xu, C., Guan, J., Zhang, H., Grieco, L.A.: Social cooperation for information-centric multimedia streaming in highway vanets. In: Proceeding of IEEE International Symposium on a World of Wireless, Mobile and Multimedia Networks 2014, pp. 1–6. IEEE (2014)
45. Arora, K., Rao, D.: Web cache page replacement by using lRU and lFU algorithms with hit ratio: a case unification. Int. J. Comput. Sci. Inf. Technol **5**(3), 3232–3235 (2014)
46. Shah, K., Mitra, A., Matani, D.: An o (1) algorithm for implementing the lFU cache eviction scheme. no 1, 1–8 (2010)
47. Gallo, M., Kauffmann, B., Muscariello, L., Simonian, A., Tanguy, C.: Performance evaluation of the random replacement policy for networks of caches. Perform. Eval. **72**, 16–36 (2014)
48. hua Ran, J., Lv, N., Zhang, D., yuan Ma, Y., yong Xie, Z.: On performance of cache policies in named data networking. In: 2013 International Conference on Advanced Computer Science and Electronics Information (ICACSEI 2013). Atlantis Press (2013)
49. Khelifi, H., Luo, S., Nour, B., Shah, C.S.: Security and privacy issues in vehicular named data networks: an overview. Mob. Inf. Syst. **2018**, 1–11 (2018)
50. Bouk, S.H., Ahmed, S.H., Hussain, R., Eun, Y.: Named data networking's intrinsic cyber-resilience for vehicular CPS. IEEE Access **6**, 60570–60585 (2018)
51. Rezaeifar, Z., Wang, J., Oh, H.: A trust-based method for mitigating cache poisoning in name data networking. J. Netw. Comput. Appl. **104**, 117–132 (2018)
52. Khelifi, H., Luo, S., Nour, B., Moungla, H., Ahmed, S.H.: Reputation-based blockchain for secure NDN caching in vehicular networks. In: IEEE Conference on Standards for Communications and Networking (CSCN), pp. 1–6. IEEE (2018)
53. Chowdhury, M., Gawande, A., Wang, L.: Secure information sharing among autonomous vehicles in NDN. In: IEEE/ACM Second International Conference on Internet-of-Things Design and Implementation (IoTDI), pp. 15–26. IEEE (2017)
54. Xia, F., Liu, L., Li, J., Ma, J., Vasilakos, A.V.: Socially aware networking: a survey. IEEE Syst. J. **9**(3), 904–921 (2015)
55. Bernardini, C., Silverston, T., Festor, O.: Socially-aware caching strategy for content centric networking. In: 2014 IFIP Networking Conference, pp. 1–9. IEEE (2014)

Internet of Things Smart Home Ecosystem

Hamza Zemrane, Youssef Baddi and Abderrahim Hasbi

Abstract Internet of things is perceived today as a key opportunity for commercial development in different sectors of activities, the new IoT Smart home ecosystem, is based on a lot of sensors and actuators that can make life in the home more pleasant and more safety for its occupant, in this article we distinguish the communication protocols used in this IoT ecosystem, their modes of operation, and we analyze their performance.

Keywords Smart home · Internet of Things (IoT) · Asynchronous Transfer Mode (ATM) · Wireless Local Area Network (WLAN) · WiFi

1 Introduction

The emergence of the connected objects that know our world today, make the Internet of Things an important factor to improve and optimize many sectors of activities. Therefore a lot new IoT ecosystems interconnect and coexist to create a smart environment that facilitate human life. In our article we focus on the new IoT Smart home ecosystem known also under the name of IoT Home automation ecosystem, witch is based on a lot sensors installed in the home that collect the home information and send it to the programmed part installed on a remote server in to Internet. This programmed part processing system can analyze the sensors information and if its detect a not normal event such as a broken window or a lot of smoke or even an expired food, it can active the specific actuator like the alarm of the house, the gas vacuum or

H. Zemrane (✉) · A. Hasbi
Mohammadia School of Engineers (EMI), Rabat, Morocco
e-mail: zemranehamza93@gmail.com

A. Hasbi
e-mail: ahasbi@gmail.com

Y. Baddi
National school For Computer Science (ENSIAS), Rabat, Morocco
e-mail: baddi.y@ucd.ac.ma

© Springer Nature Switzerland AG 2020
M. Elhoseny and A. E. Hassanien (eds.), *Emerging Technologies for Connected Internet of Vehicles and Intelligent Transportation System Networks*, Studies in Systems, Decision and Control 242, https://doi.org/10.1007/978-3-030-22773-9_8

indicator of the missing food in the refrigerator, the programmed part can also warn the user on his laptop or on his smart phone. In this article we give in the section two: IoT definition, its architecture and its domain of applications, in the section three: we define the IoT Smart home ecosystem, its advantages, and how automated systems in the home work, in the section four: we define the IEEE 802.11 standard, the WiFi operation modes, transmission technologies, and its physical frames, in the section five: we define the ATM network, the ATM flow, ATM cells composition, the connection, and the ATM layers comparing to the OSI model, in the section six: we make a simulation of our IoT Smart home ecosystem with the OPNET Network simulator, and we analyse the performance of the ATM core network, the performance of wireless local area network the WiFi protocol, and we compare the performance of the core network ATM and IP in case of communicating with the database, and in the case of using a voice application by the sensors and the actuators.

2 Internet of Things (IoT)

2.1 Definition of IoT

We talk about the Internet of Things when the number of devices exceeds the number of people connected to the Internet, the goal of the Internet of Things is to facilitate human life by building a smart environment using smart objects that can autonomously generate data from the environment in which they are deployed and transmit this data to the Internet for decision-making. The IoT devices are usually wireless sensors [1], smart phones, RFID [2], smart homes [3], and others connected to the Internet via a plug-in connection module in a clever environment. These devices are used to collect information from the physical environment, and send it to the network edge for further processing. These devices are deployed with a network architecture and a separate data processing application according to the specific task in a particular area.

2.2 Architecture of IoT

The architecture of the IoT [4] consists of the following layers:

2.2.1 Perception Layer

It is composed of physical objects that have the ability to capture physical quantities (heat, humidity, vibration, radiation, and others) and transform them into digital magnitudes, process this information, store it and transmit it via a wireless transmission module to a sink or a network gateway. This layer consists of Wireless sensors [1], RFID [2], smart phones, wearable, smart cars [5], smart homes, and others.

2.2.2 Network Layer

It transmits the digital information collect from the perception layer in analog format to a sink or the network gateway for further processing on this information. In this context we find a lot of technology on constant evolution as: Low Energy Bluetooth [6], LoRaWAN [7], WiFi [8], ZigBee [9] and others.

2.2.3 Middle-ware Layer

Several IoT devices in the same domain communicate with the same compatible device, this layer makes possible to extract the information sent from different hardware equipment, to translate it into a service information, for addressing, and denomination of the requested service.

2.2.4 Application Layer

It serves as an interface for the user to access to the collected information from the perception layer and to manipulate them according to the demand of the specific domain and process them in a processing system (Fig. 1).

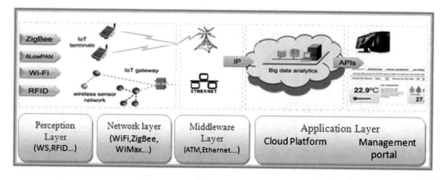

Fig. 1 Architecture of the Internet of Things

2.3   Domains of Applications of the Internet of Things

There are a variety of application areas for the Internet of Things sectors, whether in the industrial world or in everyday life.

- **Medical applications** [10]: In the field of health, connected objects can serve several purposes. For example, they can be used as a monitoring device in the form of a connected bracelet or connected watch which will make it possible to follow the physical activities of a patient at any time by informing for irregular signs. For people who take drugs connected objects can also inform when it's time to take the drugs. In addition, this could possibly inform the doctor of an emergency situation allowing him to locate the patient through the connected device.
- **Connected car or automobile 4.0** [11]: Driving a car has always been a place of strong convergence between safety, comfort and pleasure. The Internet of Things (IoT) is profoundly changing the automotive industry by offering new features to users. Hundreds of in-vehicle sensors will produce business-driven data to improve the user experience: your vehicle will be able to provide you with relevant advertising and content. Insurance is the first to take advantage of tracking devices by offering preferential offers according to the user's behavior. Its data may also be shared with other vehicles: your car may change your route if a "bad" driver is on your route. The Google car, autonomous car, is already in circulation in summer 2015 in the United States. The benefits and opportunities of IoT in the automotive industry are considerable. Ultimately, this would reduce accidents and traffic flow, manage CO_2 emissions and new marketing and commercial experiences.
- **Well-being and comfort** [12]: Home automation or smart home [3]. Imagine for a moment that your thermostat is able to go on its own depending on the location of your car allowing you to warm up once you get home. Also, imagine that your fridge is informing you when you need to buy milk or that it can create a personalized shopping list based on your most purchased items. Or tell you when your food is about to expire.

3   IoT Smart Home Ecosystem

The origin of the word home automation means a set of processes dedicated to the reduction (or elimination) of the effort and intervention of the person in the performance of a function. In the field of residential home automation, equipment are used to automate applications relating to: occupant safety, their communications, energy management (lighting and heating control), multimedia entertainment and others. We can therefore distinguish two areas of application of home automation:

- **The management of energy flows**: water, gas, electricity, when talking about domestic functions such as heating, lighting, ventilation, control of household appliances.

– **Data flow management**: telephone, radio and television, computer.

This brain dedicated to the management of the house makes life easier and more enjoyable for its occupants. In addition, intelligent heating and lighting management can have a significant impact on energy conservation without compromising the comfort of the home.

3.1 Domestic Comfort

3.1.1 A House That Simplifies Everyday Life

All actions performed mechanically can be automated and integrated into pre-programmed scenarios. Eliminating tedious and repetitive gestures that can save us time and peace of mind.

– **The home speaks to us**: Transmissions of information via mobile phone, desktop PC or laptop by a voice message, an e-mail or SMS.
– **We talk at the home**: It can be to control the voice lighting circuits or to trigger scenarios. We can activate at a distance the heating, the alarm, the shutters, etc ...

3.1.2 Autonomous House

The smart home allows a self supervision of the house system and reaction in case of need.

– **Supervision**: An autonomous house must be able to detect the changes of state of the systems to be monitored, in particular: breakdown of household appliances, malfunction of the heating system, change of weather conditions, power failure, attempted intrusion, risk household (leakage of water or gas, for example).
– **Reactivity**: examples of reactivity of the systems: The banana awning rolls thanks to a wind or rain sensor. The shutters, on the south side, go down when it is too hot. The water supply is shut off automatically by a solenoid valve in case the watering system is triggered by the information provided by humidity sensors or a local weather forecasting system via the Internet, the intensity of the lighting adapts to the external brightness.

3.2 Home Automation Systems Operation

We distinguishes in every home automated system two parts which are: the information chain and the energy chain.

- **Information chain**: to operate, automatic system must be able to acquire set-points from the user, but also the system itself and its environment, process these information and transmit orders to the energy chain.
- **Energy chain**: An automatic system must be supplied with energy to achieve its functions. The orders coming from the information chain lead to distribute the energy, to convert it and finally to transmit it (Fig. 2).

3.2.1 Example: The Central Heating of a House

A central heating system provides warmth to the whole interior of a building (or portion of a building) from one point to multiple rooms. When combined with other systems in order to control the building climate, the whole system may be an HVAC (heating, ventilation and air conditioning) central system (Fig. 3).

The central heating management of a smart home [3] can save a lot of energy, if every body live the home it stops automatically, it also allows programming heating limits, and we can connect the heater to the right sensors, this allow us to save even more energy. The information chain takes information from the user, the house and

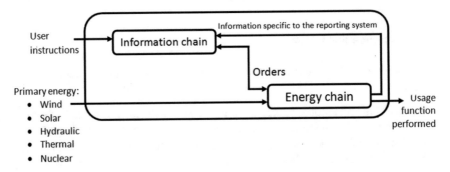

Fig. 2 Diagram of operation of an automated system

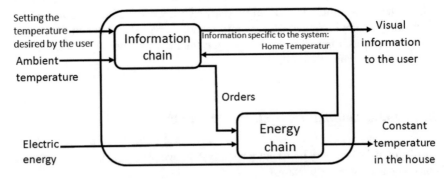

Fig. 3 Diagram of operation of the central heating of a house

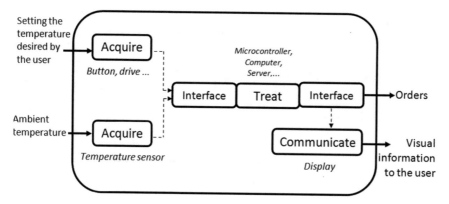

Fig. 4 Diagram of the information chain

the heating system it self and can inform of the user of the temperature, in the other hand, and the information chain send orders to the energy chain witch is also powered by an electric energy thus to create a constant temperature in the house.

Description of the information chain diagram:

– The information acquired via the temperature sensor, the button or the drive is sent directly as a signal based on a wireless communication protocol [13] at the input interface of the processing unit. Input interface is often used to adapt the signal.
– This processing unit can be a micro-controller, a local computer or a remote server.
– After the processing of the information by the algorithms designed by the operator, the result is flooded by the output interface of the processing unit using a wireless communication protocol [13], in the one hand, in the form of order to the energy chain, in the other hand, to communicate it to the user using an actuator [14] (display, speaker or indicator).
– The communication between the Acquire part, the interfaces of the processing unit, the communicating part and the distributing part of the energy chain is made alive often with wireless communication protocols [13]. We distinguishes a WLAN protocol.

Wireless Local Area Networks (WLANs): spans a relatively small area such as a building or a group of buildings, the most modern WLANs are based on IEEE 802.11 standards and marketed under the name of WiFi [8]. WiFi: is a wireless communication technology for wireless local area networks with a high data rate and band with and a large coverage area (Fig. 4).

Description of The energy chain diagram:

– It starts with a power box that receives input energy (primary energy: wind, solar, nuclear, …) and produces an output energy: often via EDF in 230v or 400v which can be either an energy electric, pneumatic, hydraulic, or other.
– The distributing part allows the piloting of power energy. "Power tap" for example: Contactor, solenoid valve, pneumatic distributor, relay, transistor, or other. This allows to have a controlled energy.

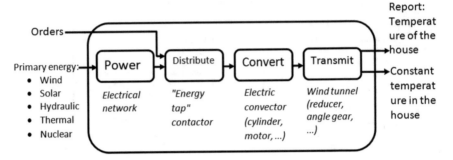

Fig. 5 Diagram of the energy chain

– The convert part makes it possible to transform the controlled energy by using a motor, a cylinder, a heating resistor, or other.
– The transmit part allows you to manipulate the transforming energy using a bevel gear, a reducer, or other (Fig. 5).

4 Network Layer for the IoT Smart Home Ecosystem: WiFi Protocol

In our simulation we will use the WiFi protocol [8] witch is a IEEE 802.11 standard to allow access at the collected information by the sensors to the remote server installed in the Internet, and to receive the result of the programmed part by the actuators [14]. The IEEE 802.11 standard defines the first two low layers of the OSI model, namely the physical layer and the data link layer. The latter is itself subdivided into two sub-layers, the LLC sublayer (Logical Link Control) and the MAC sublayer (Medium Access Control). The following figure illustrates the model architecture proposed by the 802.11 work-group compared to the OSI model. One of the peculiarities of this standard is that it offers several variants at the physical level, while the link part is unified (Fig. 6).

OSI Layer 2 *Data Link Layer*	802.11 Logical Link Control (LLC)					
	802.11 Medium Access Control (MAC)					
OSI Layer 1 *Physical Layer* *(PHY)*	FHSS	DSSS	IR	Wi-Fi 802.11b	Wi-Fi 802.11g	Wi-Fi5 802.11a

Fig. 6 WiFi layers

4.1 Typologies and Operating Mode

There are two modes of operation for WiFi:

4.1.1 Infrastructure Mode

is based on a special station called Access Point (AP). This mode allows WiFi stations to connect to a network via an access point. It allows a WiFi station to connect to another WiFi station via their common AP. A WiFi station associated with another AP can also interconnect. All AP radio range stations form a Basic Service Set (BSS). Each BBS is identified by a 6-byte BSSID (BSS Identifier) [15] that corresponds to the MAC address of the AP (Fig. 7).

4.1.2 Ad-Hoc Mode

the client wireless machines connect to each other in order to constitute a point-to-point network (peer to peer), that is to say a network in which each machine plays at the same time the role client and the role of access point. The set formed by the various stations is called a set of independent basic service set (IBSS). In an ad hoc network, the scope of the independent BSS is determined by the scope of each station. This means that if two of the network stations are out of reach of each other, they will not be able to communicate even if they "see" other stations (Fig. 8).

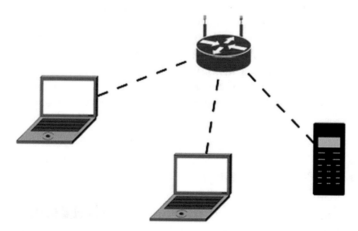

Fig. 7 Infrastructure topology

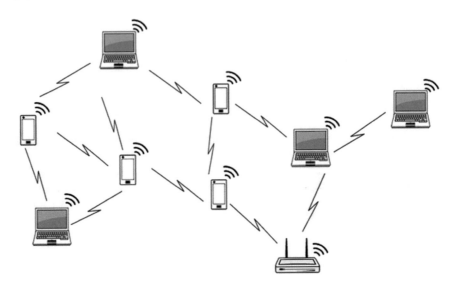

Fig. 8 Ad-hoc topology

4.2   The Transmission Channels

The transmission channel is a narrow frequency band that can be used for communication. The governments propose frequency bands for free use. The bodies responsible for regulating the use of radio frequencies are: ETSI (European Telecommunications Standards Institute) in Europe, the Federal Communications Commission (FCC) in the United States, and MKK (Kensa-kentei Kyokai) in Japan.

United States released three frequency bands for Industry, Science and Medicine. These frequency bands, called ISM (Industrial, Scientific, and Medical), are the bands 902–928 MHz, 2.400–2.4835 GHz, 5.725–5.850 GHz.

In Europe the 890–915 MHz band is used for mobile communications (GSM), so only the bands 2,400–2,44835 GHz and 5,725–5,850 GHz are available for amateur radio use.

4.3   Transmission Technologies

Local radio networks use radio or infrared waves to transmit data. The technique originally used for radio transmissions is called narrow-band transmission; it consists of passing different communications over different channels. The physical layer of the 802.11 standard thus initially defines several transmission techniques to limit the problems due to interference: The technique of frequency hopping spread spectrum, the technique of direct sequence spread spectrum, and the infrared technology.

4.3.1 The Narrow-band Technique

The narrow band technique consists in using a specific radio frequency for the transmission and reception of data. The frequency band used should be as small as possible to limit interference on adjacent bands.

4.3.2 The Technique of Frequency Hopping Spread Spectrum (FHSS)

consists of splitting the wide frequency band into a minimum of 75 channels [16], then transmit using a known channel combination of all stations in the cell. In the 802.11 standard, the 2.4–2.4835 GHz frequency band makes it possible to create 79 channels of 1 MHz. The transmission is done by emitting successively on one channel then on another for a short period of time (about 400 ms), which allows a given moment to transmit a more easily recognizable signal on a given frequency.

4.3.3 The Technique of Direct Sequence Spread Spectrum (DSSS)

consist of transmitting for each bit a Barker sequence of bits [17]. The physical layer of the 802.11 standard defines an 11-bit sequence (10110111000) to represent a 1 and its complement (01001000111) to encode a 0. Each bit encoded with the sequence is called chip or chipping code.

4.3.4 Infrared Technology

The IEEE 802.11 standard also provides an alternative to using radio waves: infrared light. The main characteristic of infrared technology is the use of a light wave for data transmission. Thus the transmissions are uni-directional. The non-dissipative nature of the light waves provides a higher level of security. It is possible thanks to the infrared technology to obtain rates ranging from 1 to 2 Mbit/s using a modulation called PPM (pulse position modulation).

4.4 WiFi Physical Frames

Data packets from the network layer are encapsulated at level 2 by a MAC header, forming a Mac Protocol Data Unit (MPDU). This MPDU is then encapsulated in a second frame at level 1 (physical) to allow transmission on the media this set forms a (PLCP-PDU) (Figs. 9, 10 and 11).

The FHSS frame used in the IEEE 802.11.
DSSS frame used in the IEEE 802.11 b.
Frame OFDM used in the IEEE 802.11a and the IEEE 802.11g.

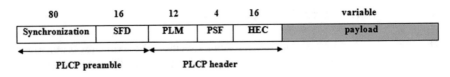

Fig. 9 FHSS frame bit size

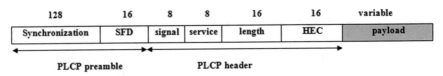

Fig. 10 DSSS frame bit size

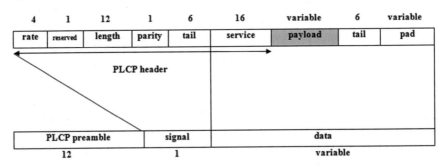

Fig. 11 OFDM frame bit size

5 Middleware Layer for the IoT Smart Home Ecosystem: ATM Protocol

ATM (Asynchronous Transfer Mode) networks [18] must allow the transport, multiplexing and demultiplexing of all types of traffic (voice, data, images) on the physical media of the operators public networks at the national or continent level. ATM [18] must allow high throughput and low switching delay.

– **In switching**: Traditional switches will be turned into a broadband switch with ATM applications.
– **In the transmission**: Use of ATM switching nodes and multiplexing/demultiplexing nodes.
– **In the Internet**: ATM routers are involved in IP switching and MPLS applications.

5.1 Nature of ATM Flow

The streams of data that must be conveyed are:
Isochronous: for the transfer of periodic data such as sound or moving image (with or without compression)

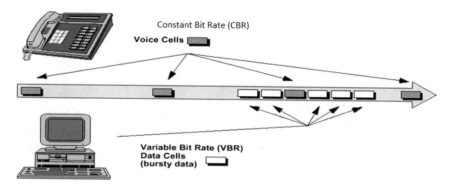

Fig. 12 Example of isochronous and asynchronous traffic

Asynchronous: with or without bursts, for the transfer of data between local networks (Fig. 12).

5.2 Principles of ATM Networks

Based on cell transmission and switching. Virtual connections and asynchronous data multiplexing (on one or more physical media). ATM networks [18] must provide each customer with a contract of:

- Flow adapted to its needs and likely to evolve at will.
- Guarantee a transport time compatible with comfort necessary for interactive applications.
- Minimal control for transmission and flow errors.
- High speed: 155 Mbps, 622 Mbps, 2.4 Gbps,…
- Compromise between circuit and packet switching.

5.3 ATM Cells

5.3.1 Size of a Cell

The size of an ATM cell is 53 bytes, whose length of the data area is 48 bytes. The length of the data area is the result of a compromise between Europeans who wanted 32 bytes and Americans who wanted 64 bytes. The very short cell length is explained by the desired quality in the transmission of a signal. In fact, for the transport of the voice, the transmission time (propagation, treatment,…) must remain less than 28 ms, to avoid the cancellation of echo and adaptation.

5.3.2 Why the Choice of 48 Bytes as the Size of the Data Area

If we look at the transit time of bytes for the speech coming from a telephone booth we have:

Sampling 1 byte/125 μs so it takes 6 ms for fill 48 bytes.

28 ms max = 6 ms (fill) + 6 ms (dump) + 16 ms.

Assuming that the signal is transmitted over an electrical cable at a speed of 200000 km/s, the maximum distance that the signal can travel without the echo being detected is 3200 km. As the US territory is very extensive, Americans opt for an extension of the cell data area compared to the supervision part.

5.4 Head of a Cell

5.4.1 ATM Cells

The header of the ATM cell is composed of:

- **GFC**: It assigned only for connections between host and network, no end-to-end significance, not delivered, to the receiver (the fields are overwritten at the first encountered router ...). In the case of unregulated equipment, the GFC function is not used. Therefore, no action is taken on the positioning of the GFC field where all bits are always set to 0 on transmission. When implemented, the GFC control procedures provide the following three functions:

  - One option is to cyclically stop (HALT) the transmission of traffic on all ATM connections in order to limit ATM traffic [18] to the network through the UNI (user network interface) at a fixed fraction of the interface throughput.
  - Network access control for traffic on regulated ATM connections.
  - The explicit indication, from the regulated equipment to the control equipment, that a cell is offered over a regulated ATM connection.

- **VPI**: Virtual Path Indicator.
- **VCI**: Virtual Circuit Indicator, a VP contains several VCs.
- **PTI**: It indicates the data types of the payload, and allows the receiver to indicate the level of congestion.
- **CLP**: It indicates the level of priority, ATM switches drop a cell with CLP = 1 before a cell with CLP = 0.
- **HEC**: 8-bit CRC, with the generator polynomial

$$g(x) = x^8 + x^2 + x + 1$$

It can correct all errors on 1 bit, and it can detect 90% of all other types of errors (Fig. 13).

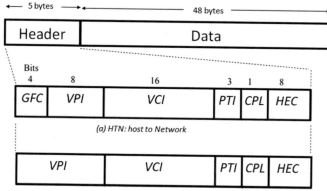

Fig. 13 Header of an ATM cell

Fig. 14 ATM physical link

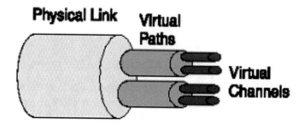

5.5 ATM Connections

Between two network interfaces, a connection (called VC for Virtual Circuit), is identified by a number, the VCN (Virtual Connection Number). The VCN breaks down into two other numbers: its high-order bits make up the VPI (Virtual Path Identifier), and its least significant bits the VCI (Virtual Channel Identifier). A VP is a bundle of VC. Channel identification in the ATM cell header: VCI : Virtual Channel Identifier, and VPI: Virtual Path Identifier (Fig. 14).

ATM [18] is connection-oriented, so that all packets traverse a virtual circuit. An ATM virtual circuit is a logical connection from a source to a destination; it is also possible to establish multi-cast connections. A CV is unidirectional, but a pair of CVs can be created at the same time between two points (one in each direction). A circuit signaling can be:

- **Permanent Virtual Circuit (PVC)**: The connections are analogous to the dedicated lines that are switched between certain users. A change can only be made by the operator.
- **Switch Virtual Circuit (SVC)**: Users who are connected to this type of network can establish a connection of their own choice via the signaling procedures. This can be compared to the call setup (Fig. 15).

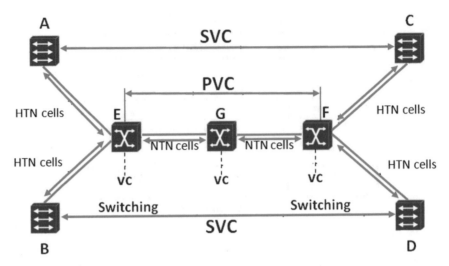

Fig. 15 PVC and SVC signaling

For the connection in ATM there are two types:

- **VCC (virtual channel connection)**: Defines the end-to-end connection between two access points to the AAL layer. A CCV is composed of the cancer of one or more VC.
- **VPC (virtual path connection)**: is composed of the concatenation of one or more VP. ATM node is called brewer (Fig. 16).

5.6 ATM Sublayer Model

ATM networks [18] follow the principles of a new architecture where the functionalists are not grouped at the same levels as in the reference model (Fig. 17).

5.6.1 The Physical Layer

The physical layer of this new model corresponds to the physical layer of the OSI model, but with one important difference: It groups the bits by 424 to directly find the structure of the cell. It transmits on the physical medium: voltage, bit sampling, etc. It is divided into two sub-layers, the lower sub-layer PMD (Physical Medium Dependent) and the under layer (TC) (Transmission Convergence).

PMD sub-layer The PMD sub-layer interfaces with the physical medium. It sets the bits to 1 or 0 and manages their synchronization. It differs from one operator to another or from one carrier to another. PMD describes how cells are emitted on the physical medium.

(a)

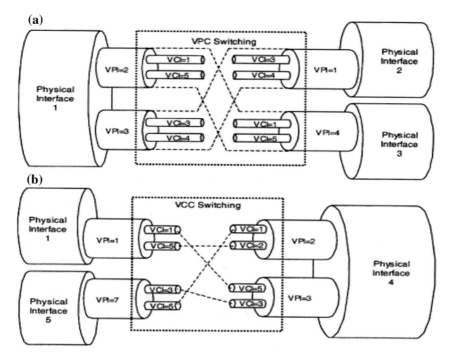

(b)

Fig. 16 VCC and VPC connection

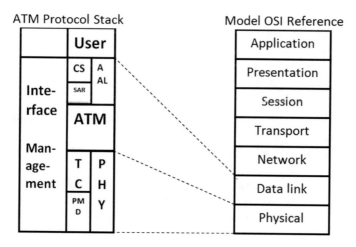

Fig. 17 ATM comparison with OSI reference

TC sub-layer Allows error detection on header only.
Header = 4 bytes of address + 1 byte CRC (HEC)

$$g(x) = x8 + x2 + x + 1$$

It adapts to the flow of the physical layer. Framing at the reception of the cell.

5.6.2   The ATM Layer

This layer provides the following functions: Multiplexing and demultiplexing, virtual address translation, flow control, cell switching, processing the header, quality of service, determining the type of the payload. It is connection-oriented, based on virtual circuits.

5.6.3   AAL Layer

AAL is an interface between higher software layers and low-level 48-byte cell-to-data protocols from one point of the network to another. Different types of AAL will therefore be used depending on the needs of the upper layers.

CS sub-layer make management transmission times, detection of lost cells, detection of transmission errors

SAR sub-layer cut CS-PDUs for implementation in ATM cells with 48 bytes of data.

Types of AAL Classes of services:

Class A: AAL 1, voice, video, CBR
Class B: AAL 2, video, VBR
Class C: AAL 3/4, data, VBR
Class D: AAL 5, data, VBR

- CBR = Constant Bit Rate: guarantees a constant bandwidth, guaranteed peak traffic: real-time video, telephony.
- VBR = Bit Rate Variable: Ensures average throughput, all data is transmitted, but can be delayed (delay not guaranteed).
- UBR = Unspecified Bit Rate: Without any guarantee (Best Effort).
- ABR = Available Bit Rate: use of the available bandwidth without loss of information, indication of congestion by the CLP bit = 1 (Table 1 and Fig. 18).

Table 1 AAL types of Service

Type of Service	A	B	C	D
Bit rate	Constant	Variable	Variable	Variable
Real time	Yes	Yes	No	No
Connection mode	Oriented	Oriented	Oriented	Without
AAL	AAL1	AAL2	AAL3/4	AAL3/4,AAL5

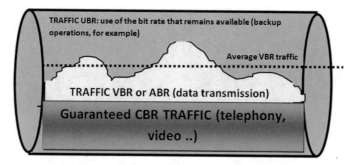

Fig. 18 Distribution of traffic in an ATM conduit

6 Simulation of the IoT Smart Home Ecosystem Using OPNET Network Simulator

Our IoT Smart home ecosystem is based:

– On the perception layer of the architecture of the IoT ecosystem on four sensors they can be a glass breaking sensor, garage door sensor, central heating sensor of the house, or a smoking sensor, that can send the information collected from the home to the Internet for further processing.
– On the network layer of the architecture of the of the IoT ecosystem we used the IEEE 802.11 protocol marked under the name of WiFi [8] to collect the environmental information from every sensor and send it to the network getaway.
– On the middleware layer of the architecture we used the ATM protocol [18] and we compare it to the IP protocol witch still the unique protocol used in the interconnection between the Internet networks.
– In the application layer we used a processing system dedicated to the Smart Home [3] host on a WEB server in the Internet.

When the sensors are configured they can send every second (or less) the collected information form the environment to the processing system located in the Internet, which allows the user to know the state of his house even if he is out side, using his laptop or smart phone.

And if the processing system detects a not normal event in the house such as breaking a window, opening a door, a lot of smoke, or a very high temperature, the processing system send a message to active the special actuator [14], such as activating the house alarm, the smoke aspirator, or the triggering of the water, and warn the user on his laptop or smart phone. The simulation is done by the OPNET Network simulator, the 14 January 2019, it lasts for 40 minute it begun at 05:53:20 pm and ends at 06:36:40 pm (Figs. 19 and 20).

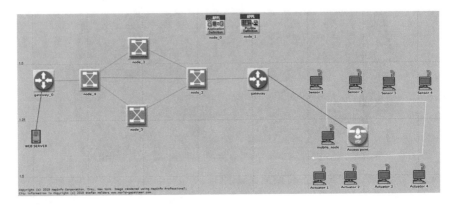

Fig. 19 Simulation scenario based on the core network ATM

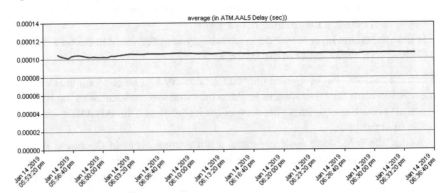

Fig. 20 Simulation scenario based on the core network IP

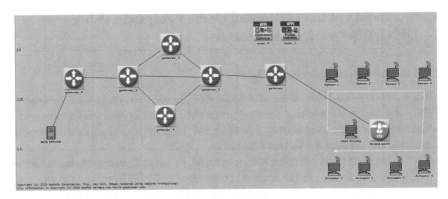

Fig. 21 Average of the AAL5 delay

6.1 Performance Study of the ATM Porotocol

AAL5 delay is the amount of delay from input to output of a logic gate in a path. This delay is caused by the parasitic capacitance of the interconnection between the two cells, combined with net resistance and the limited drive strength of the cell driving the net. When the simulation starts the curve is just over 0.0001 s and remains more or less constant on this value until the end of the simulation (Fig. 21).

AAL 5 delay variation is a term used in ATM to describe the time difference that is acceptable between the AAL5 cells being presented at the receiving host. The curve starts with 0.8 μs after it begun to decrease to reach 0.5 μs at 06:06:00 pm after that it starts to increase to stop at 0.67 μs near the end of the simulation (Fig. 22).

6.2 Performance Study of the WiFi Porotocol

Delay specifies how long it takes for a bit of data to travel across the network from one node or endpoint to another. The simulation start with 0.255 s then the curve tries to stabilise until the end of the simulation to reach 0.30 s (Fig. 23).

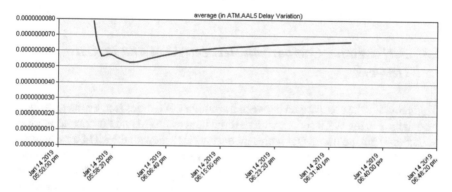

Fig. 22 Average of the AAL5 delay variation

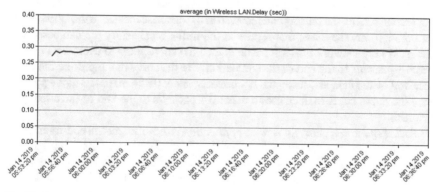

Fig. 23 Average of the WLAN delay

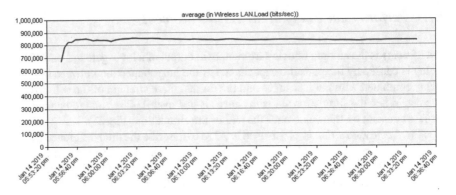

Fig. 24 Average of the WLAN load

Load it's the number of frames that transit over the wireless network as a function of time. The curve starts with a little bit less than 7,00,000 bit/s, then it starts to increase to reach up to 8,00,000 bit/s at 05:56:40 pm then the curve try to stabilise at this value until the end of the simulation (Fig. 24).

6.3 Performance Comparison Between ATM and IP Core Network

Traffic sent it's the comparison between the number of queries sent by the sensors, to the database in the WEB server, with the core network ATM and the core network IP. At 05:56:40 pm the traffic that represent the queries sent with the ATM core network is at 4.5 queries/s and then it starts to decrease slowly until the end of the simulation when it reach 3 queries/s. for the curve that represent the queries sent with the IP core network at 05:56:40 pm the curve is at 3.5 queries/s then starts to decrease slowly to reach 2.3 queries/s to the end of the simulation (Fig. 25).

Traffic received by the database it's the comparison between the number of queries per second sent by the sensors and received to the database located in the WEB server. At 05:56:40 pm the curve that represent the traffic received by the database using the ATM core network is at 1.3 packet/s then the curve starts to decrease slowly to reach 1 packet/s to the end of the simulation. For the curve that represent the traffic received by the database using the IP core network at 05:56:40 pm its at 0.7 packet/s then the curve starts to decrease to reach 0.4 packet/s to the end of the simulation (Fig. 26).

Packet end to end delay for a voice application is the difference in end-to-end one-way delay between selected packets in a flow with any lost packets being ignored, for a voice application used by the sensors and the actuators [14]. For the cure based on the ATM core network it starts just over 0.4 s then it tries to stabilise on this value

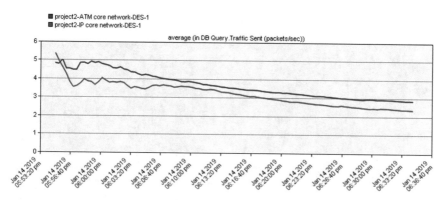

Fig. 25 Average of the queries sent by the sensors

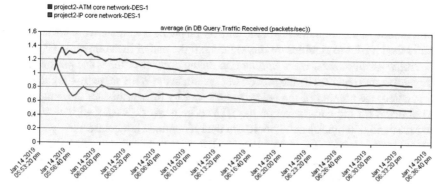

Fig. 26 Average of the queries received by the database

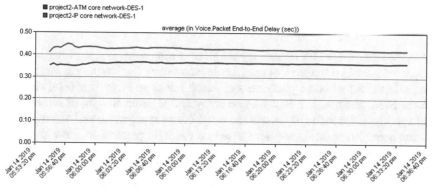

Fig. 27 Average of the voice packet end to end delay

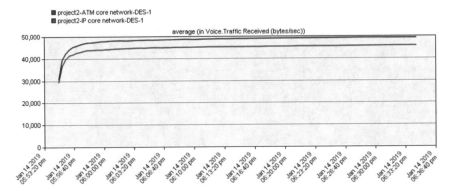

Fig. 28 Average of the voice traffic received

until the end of the simulation. For the cure based on the IP core network it starts with 0.36 s then it tries to stabilise on this value until the end of the simulation (Fig. 27).

Traffic received for a voice application its the number of the voice packets from the sensors and received by the processing system in the WEB server to the end of the simulation, the number of the voice packets sent by the sensors are the same using ATM or IP. for the cure based on the ATM core network it stars with an increase that reach 45,000 bytes/s at 05:56:40 pm then it continue it increase slowly to reach 50,000 bytes/s toward the end of the simulation. For the cure based on the IP core network it stars with 30,000 bytes/s after it increase rapidly to have 42,000 bytes/s then increase slowly to reach 45,000 bytes/s at the end of the simulation (Fig. 28).

7 Conclusion

Over the years, the field of technology has given us a impressive range of progress and improvements. One of the most remarkable innovations that has the greatest impact so far is the Smart home phenomenon, which brings so many beneficial changes to our lives. The new IoT Smart home ecosystem is based on a lot of sensors and actuators installed in the home, this can make life more enjoyable and secure. In our Smart home IoT solution we were based on the WiFi protocol to allow access at the collected information of the house at the Web server located in the Internet that contain the programmed part of our Smart home, and to receive the orders that will be executed by the actuators. For core network we were based on the ATM protocol network, and we compare it to the IP protocol network that still the unique protocol used in the interconnection of the Internet networks.

References

1. Abbasi, A.Z., Islam, N., Shaikh, Z.A., et al.: A review of wireless sensors and networks' applications in agriculture. Comput. Stand. Interfaces **36**(2), 263–270 (2014)
2. Zhu, X., Mukhopadhyay, S.K., Kurata, H.: A review of RFID technology and its managerial applications in different industries. J. Eng. Technol. Manag. **29**(1), 152–167 (2012)
3. Sivaraman, V., Gharakheili, H.H., Vishwanath, A., et al.: Network-level security and privacy control for smart-home IoT devices. In: 2015 IEEE 11th International Conference on Wireless and Mobile Computing, Networking and Communications (WiMob), pp. 163–167. IEEE (2015)
4. Datta, S.K., Bonnet, C., Nikaein, N.: An IoT gateway centric architecture to provide novel M2M services. In: 2014 IEEE World Forum on Internet of Things (WF-IoT), pp. 514–519 (2014)
5. Mammeri, A., Zhou, D., Boukerche, A., et al.: An efficient animal detection system for smart cars using cascaded classifiers. In: 2014 IEEE International Conference on Communications (ICC), pp. 1854–1859. IEEE (2014)
6. Faragher, R., Harle, R.: An analysis of the accuracy of bluetooth low energy for indoor positioning applications. In: Proceedings of the 27th International Technical Meeting of the Satellite Division of the Institute of Navigation (ION GNSS+14), pp. 201–210 (2014)
7. Wixted, A.J., Kinnaird, P., Larijani, H., et al.: Evaluation of LoRa and LoRaWAN for wireless sensor networks. In: 2016 IEEE SENSORS, pp. 1–3. IEEE (2016)
8. Duarte, M., Sabharwal, A., Aggarwal, V., et al.: Design and characterization of a full-duplex multiantenna system for WiFi networks. IEEE Trans. Veh. Technol. **63**(3), pp. 1160–1177 (2014)
9. Malhi, K., Mukhopadhyay, S.C., Schnepper, J., et al.: A zigbee-based wearable physiological parameters monitoring system. IEEE Sens. J. **12**(3), pp. 423–430 (2012)
10. Zhang, Y., Sun, L., Song, H., et al. Ubiquitous WSN for healthcare: recent advances and future prospects. IEEE Internet Things J. **1**(4), pp. 311–318 (2014)
11. Coppola, R., Morisio, M.: Connected car: technologies, issues, future trends. ACM Comput. Surv. (CSUR) **49**(3), p. 46 (2016)
12. Wang, J.-C., Lin, C.-H., Siahaan, E., et al.: Mixed sound event verification on wireless sensor network for home automation. IEEE Trans. Ind. Inform. **10**(1), pp. 803–812 (2014)
13. Rosenblatt, M.: Personal media devices with wireless communication. U.S. Patent No 8,718,620, 6 May 2014
14. Ionov, L.: Hydrogel-based actuators: possibilities and limitations. Mater. Today **17**(10), pp. 494–503 (2014)
15. Sakib, M.N., Halim, J.B., Huang, C.T.: Determining location and movement pattern using anonymized wifi access point bssid. In: 2014 7th International Conference on Security Technology, pp. 11–14 (2014)
16. Hughes, S., van Greunen, J., Vaswani, R., et al.: Creation and use of unique hopping sequences in a frequency-hopping spread spectrum (FHSS) wireless communications network. U.S. Patent No 8,442,092, 14 May 2013
17. Spuhler, M., Giustiniano, D., Lenders, V., et al.: Detection of reactive jamming in DSSS-based wireless communications. IEEE Trans. Wirel. Commun. **13**(3), pp. 1593–1603 (2014)
18. Nordstrom, E.: A hybrid admission control scheme for broadband ATM traffic. In: Proceedings of the International Workshop on Applications of Neural Networks to Telecommunications, pp. 87–94. Psychology Press (2013)

An Enhanced Adhoc Approach Based on Active Help to Detect Data Flow Anomalies in a Loop of a Business Modeling

Najat Chadli, M. I. Kabbaj and Z. Bakkoury

Abstract Data flow in business process modeling is created and distributed by the exchange of data moving from one task to another in information systems. Among open issues in workflow modeling is the detection of errors in data flow and control flow. Researchers have recently focused on detecting errors by applying an active help with a concept of Data-Record. However, this method does not support a loop modeling. This chapter presents a novel active help with a Data-Record concept in order to detect data flow anomalies in loop modeling. We the improve the active help approach using suitable rules for loop modeling where a decision node, using a data connection as an input data, replaces the connector Xor-split. The input data of the decision node is returned to the last activity as a feedback when the error message is found. The proposed approach is validated using a deterministic finite state process model which uses a logic Boolean predicate (Yes or No) to specify the routing of an input data. In this chapter, anomalies such as missing data, conflicting data and redundant data are investigated. The verification is triggered when an anomaly is detected, and the system is locked until a correction is performed. The results show that Missing Data anomalies are efficiently handled by the proposed approach when there is a feedback loop. The simulation is carried out using The Drools platform which introduced RuleFlow tool in KIE (Knowledge Is Everything). For the other anomalies, conflicting data and redundant data are also verified by a uppaal tool in model checking.

Keywords Data flow modeling · Active help · Loop modeling · Verification · Validation

N. Chadli (✉) · M. I. Kabbaj · Z. Bakkoury
AMIPS Research Group, Ecole Mohammadia D'Ingénieurs, Mohammed V University, Rabat, Morocco
e-mail: najatchadli@research.emi.ac.ma

© Springer Nature Switzerland AG 2020
M. Elhoseny and A. E. Hassanien (eds.), *Emerging Technologies for Connected Internet of Vehicles and Intelligent Transportation System Networks*, Studies in Systems, Decision and Control 242, https://doi.org/10.1007/978-3-030-22773-9_9

127

1 Introduction

Currently, many business functions such as purchasing, manufacturing, marketing, engineering, and accounting have been automated by most organizations [1]. To use these functions, data exchange in an information system is necessary from a task to the next in the business process management [2]. In this sense, each task requires input and output data flow [3]. Specific collection of tasks, resources and information elements make up a workflow system instance [1]. Indeed, business process activities are realized throughout tasks in the information systems. Also, business process activities can be achieved by information systems without any human involvement [4]. In fact, in business process management, it is necessary to use a workflow system [5] in order to interplay between data flow created by a data exchange of information systems and the control flow in a workflow [6]. Consequently, data flow is important for business process integration because data is always classified when conducting inter-organizational business and data errors could still happen even in the case of syntactically correct activity dependence [7, 8]. However, the majority of existing approaches focuses on control perspective and data perspective to describe the logical order of tasks and the information exchange between tasks on verification, i.e., on the discovery of design errors. Certainly, the flow-oriented nature of workflow processes like the Petri net formalism is a natural contender for modeling and analysis of workflows [9, 10]. For correctness of business processes, both control flow and data flow errors have to be considered [11]. Many solutions have been proposed to solve the data flow anomalies, as each activity needs operational information to define the state of data, that is Read, Write or Destroy. Therefore, this operation can specify the state of data in the activity to another that can cause missing data, conflicting data and redundant data [12]. An ad hoc approach which uses active help [13] is proposed in a linear model and Xor-split with two branches. This approach helps to verify the task and correct errors at the same time using a locked system and the concept of DataRecord [13]. Moreover, they do not use the loop when the ad hoc system [14] has a problem in a sent message in a linear model or model with xor branches. In this way, the same approach is used when adding a loop modeling in the linear model with an Xor split in order to detect data flow modeling anomalies. Indeed, the Xor-split is used to feedback an existing message errors at the start of modeling, this error message is returned to the source activity where it was created up to proceed to a correction. Therefore, using active help [13], and the rules for model verification, that is triggered when some issue has occurred in the time of modeling. However, the loop couldn't adapt this approach to detect anomalies not because the active help is insufficient but because the rules of this approach could only create and update. Subsequently, it's proposed to improve this approach with some enhancements in rules and model in order for the approach to be adapted to loops. A decision node is proposed like a connector that has a data connection at the input data. In this case, it requires a Boolean predicate (Yes = true, No = false) in a determinist finite-state automaton, so we used the guarding (i.e. blocking) tasks solely on the DataState [15]. In this context, we implemented Data-State to verify the last record state of the dataset for each input and output in the activity. In this manner, this data connection is a decision variable that is a routing decision that can be made based on a set of data items inputted to the decision node. Each of these data items involved in a routing decision is called a decision variable [12]. Moreover, this decision variable is allowed to change the state of Data-State that can be initialized in each iteration of the connection. Also, there is no problem in the first iteration; however, when the number of iterations is high, the initialization of the DataState is required.

The remaining of the chapter is organized as follows. Section 2 presents some approaches and concepts used in this chapter. Section 3 shows that the loop modeling cannot integrate the approach with active help. Section 4 presents the new approach visualization and the implementation of missing data rule, conflicting data rule, and redundant data rule. Section 5 concludes the chapter and discusses the perspective.

2 Related Work

Modeling in the business process has become very important in recent years, with data-flow modelling and verification being the two important challenges in workflow system management. There many many stakeholders in this problem of anomalies of data-flow and control flow in workflow systems. Recently, data flow formalization in process modeling has been investigated by many researchers. In most organisations, it is particularly important that people in charge of key processes feel their interests are represented during the latter phase. To achieve this, the main stakeholders such as the heads of key functions intersected by the process, the managers with operational responsibility for the process, suppliers of important change resources (e.g., the IT, human resource, and financial functions), and process customers and suppliers, both internal and external should participate in the team during the design phase [4]. Certainly, data flow perspective approach formally discovers the correctness criteria for data-flow modelling. Petri Net based approach is proposed for modelling control flow of workflow. We extended this model by including data flow input and output and adding a complexity of algorithm for detecting data flow anomalies as in [9]. Many approaches have been proposed for data-flow verification, these approaches enable systematic and automatic elimination of data-flow errors. Although, the dataflow verification will necessarily differ from a formal program verification, the data flow errors will be detected by a set of stated correctness criteria, in spite of formal program verification that is caused uniformity of verification between a specification and related execution results as in [12]. The information perspective defines what data are expended and produced for, with reference to each activity in a business process. Thus, the operational perspective requires the tools and applications which are used to execute a particular task [12]. The importance of data-flow verification in workflow processes was first mentioned in [16]. So as a three-layer workflow model for designing a workflow was proposed in [17]. They characterized the behavior of an artifact by its state transition diagram and identified six inaccurate usages affecting workflow execution and a set of algorithms to detect these inaccurate usages in workflow specification [17]. It has, therefore, the potential to be developed into a unified algorithmic procedure for the concurrent detection of control flow and data flow errors. Sundari et al. proposed an algorithm called GTforDF, for data flow verification through the detection of lost data. They explain through practical examples how GTforDF detects data flow errors in workflows and define an important new error category called redundant data in loops that can lead to data loss in some situations [18]. Treka et al. proposed a new analysis framework that is based on (a) workflow nets with data, (b) temporal logic, and (c) anti-patterns [19]. A Workflow net with Data (WFD-net) is a special type of a Petri net, with a clear start and end point and annotations related to the handling of data so that the activity may read, write, or destroy a particular data element [19].

In graph-based approaches to business process modelling, data dependencies are represented by data flow between activities. Each process activity is given a set of input and a set of output parameters. Upon its start, an activity reads its input parameters, and upon its termination, it writes data it generates to its output parameters. These parameters can be used by follow up activities in the business process ([20] p. 100). An approach which extends and generalizes data flow verification methods has been recently proposed. It makes use of the concept of corresponding pairs lately introduced in control flow verification. The approach focuses on the discovery of data flow errors in workflows such as redundant data, lost data, missing data, mismatched data, inconsistent data, and misdirected data. To achieve this, we propose an analysis which uses "The RWD Boolean Table Technique" that is expressed in steps, to split data-flow from control flow and to create Boolean table for each data elements, and also to compare RWD Boolean table for current task and next task until it gets to the end of the workflow [21]. Thus, an approach in data flow issues proceedings for mapping BPMN to Petri-Net to provide a systematic technique of possible flows related to the data flow of business process Data flow issues and BPMN mapping to Petri Net: Road map as in [22]. Additionally, some researchers used a survey of issues and approaches that tries to argue that a new approach that balances between those extremes is needed. Indeed, a process execution (i.e. executing the control flow) needs tasks to be enabled or disabled and this is done at the data level. If data is missing or is not available when needed, the entire execution of the workflow ends [23]. Furthermore, there exists the case where the business process of a system is sound from the view of control-flow aspect but is not correct if data-flow is considered, e.g., missing data, inconsistent data and conflict data. These data errors are a big challenge for system design. Many studies focus on data-flows in workflow systems; define several data irregularities that may lead to an incorrect execution of a workflow system. To resolve the above problem, we define a workflow-Net with Data Constraints (WFDC-net). It not only formalizes the abstract data operations but in addition, considers data constraints [24]. In a nutshell, the objective of this chapter is merging the Decision Node with an input data and a logic Boolean with the aim to find a new solution to solve problems of data flow anomalies in the business process with a loop modeling in a linear model and Xor split.

3 Verification Approach with a Loop Modeling

In the case where an ad hoc method is applied in a simple linear model with a loop, the system is triggered when a transmission error message is produced. So, a feedback is structured to return the error to be corrected. Otherwise, the modeler continues to execute the next task as in Fig. 1. So, the feedback requires the verification to detect data flow anomalies in each system workflow instance. A data flow is specified with a set of business activities, the related routing constraints defined on the given business activities, an input and output data items for each business [25]. Consequently, the output data becomes a routing constraint data in an information exchange system of data flow. Indeed, at the moment the error message is returned, the approach is triggered, and the verification is tasked to detect the anomalies in each system workflow instance. As a result, in the feedback and in each activity, the output data becomes an input data in an information exchange system. Additionally, in a workflow, each activity performs a comparison operation on a data element. Thus, data operations are spontaneous, when an activity A is reading data, the item d is an input data. The same, when an activity A carries out a writing operation on d, d is the output data from A as in [12]. Furthermore, it is proposed to verify some anomalies such as missing data, conflicting data, and redundant data for loop modeling using the approach rules [13].

3.1 Description of Symbols Used in the Model Loop

Tables 1, 2 and 3 define and describe symbols and operations used in the model in Fig. 1.

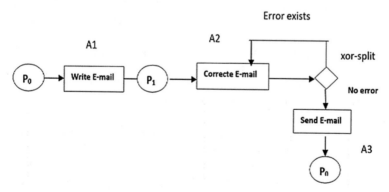

Fig. 1 Description of linear model and loop modeling

Table 1 Description of data items

Data items	Description
m	Message1(email)
m'	Message2(email)
d	Destination
e	Error
a	Accuse

Table 2 Description of activities

Activity	Description
A1	Write email
A2	Corrected email
A3	Send email
N1	Decision node

Table 3 Description of operations

Operation	Description
R	Read
W	Write
D	Destroy
1	Input
O	Output

3.2 Verification Model with a Loop

We apply missing data Rule 1 of the approach [13] in the example presented in Fig. 2.

Rule 1: "For an activity, a given data d with the state (x,y,z), if d is inserted for the first time in the DataRecord and $x \neq 0$ we have an error ==> uninitialized data (missing data)."

The data item d is detected as missing data in activity A2 in Fig. 2 and Table 5 below.

For the first activity A1, the DataRecord elements m and m' are initialized and no anomaly is detected in this state as in Table 4.

After drawing the next task for activity A2 the system detects the anomaly of data element d as a missing data. Therefore, the verification is triggered, and Rule 1 is applied. The modeler has two options to correct anomalies of missing data, either not to read data item d in activity A2 (to destroy) or write data d in activity A2 as in Fig. 2 and Table 5. Then, the modeler chooses not to read data d in activity A2. In this instance, before the first iteration, an error message e has occurred at processing in the activity A2. Consequently, after drawing the model completely and using a connector Xor-split to have the conditions for the error e to occur. This model is considered as a finite state determinist model with Xor-split as a node. Indeed, if the

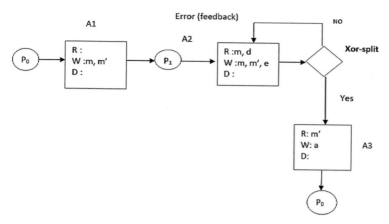

Fig. 2 Data flow modeling in linear model

Table 4 Simple state without iteration

Data	State
m	$(0, W_{A1}, 0)$
m'	$(0, W_{A1}, 0)$

Table 5 The first iteration

Data	State
m	$(R_{A2}, W_{A1}, 0)$
m'	$(0, W_{A2}, 0)$
e	$(0, W_{A2}, 0)$

error e is written in the activity A2, the feedback loop would return the message for correction, otherwise, the process continues modeling. In fact, it is required to verify the anomalies at the time of modeling to detect missing data in two cases, applying the Rule 1 of the approach [13].

If $e \neq \emptyset$: it is noteworthy that the Data-State (DataRecord) contains the latest data set, and activities as in Table 5. At the verification, the loop cannot analyse anomalies of data flow present in processing, because the loop is a feedback of the error message that occurred in the activity A2 and read what exists in this latest activity. Therefore, no missing data is detected.

Otherwise, if $e = \emptyset$: the system continues with the modeling and applies the redundant data Rules 4 in [13]. Consequently, the process verification continues to store the dataset and activities in Data-State given in Table 6, and data element a is detected as redundant data.

We conclude that for loop modeling with a connector Xor-split, even if they are based on data, this will not change anything in this approach [13], as the Rules are not practicable. Therefore, we propose to enrich this approach [13] to take into consideration the loop when error in the message occurs.

Table 6 The new dictionary with iteration

Data	State
m	$(R_{A3}, W_{A1}, 0)$
m'	$(R_{A3}, W_{A2}, 0)$
d	$(0, W_{A3}, 0)$
a	$(0, W_{A3}, 0)$

4 Description of the New Approach and Implementation

4.1 Description and Definitions

4.1.1 Data Flow Modeling

Dataflow is often defined using a model or diagram in which the entire process of data movement is mapped as it passes from one component to the next within a program or a system, taking into consideration how it changes form during the process [8]. Thus, data item of data flow has two roles; one is data link which connects an activity to another by an input data and an output data. The two roles are to transmit the information from one task to another. In our situation, data has the two roles and this data is extended by read and write/update and destroy. Consequently, when data transmit information there are many errors to be tackled such as missing data, conflicting data and redundant data.

4.1.2 Decision Nodes

The decision node is a conditional construct which can also be modeled with a conditional node using a Boolean logic predicates with a function guard that can allow us to model decision points in which the choice is made based on some data elements. When the model uses decision nodes, usually their edges have guards that are boolean logic values evaluated at runtime to discover if control and data can be evaluated along the edge. Additionally, for each individual control and data token evaluated by the guards at the decision node to get precisely the edge that the token will be extended across. We can say that decision nodes are Task nodes that represent atomic manual automated activities or subprocess that must be completed to fulfil the below business process objectives [12].

4.1.3 Data Operation

Data operation is exactly what some organization produces in their day to day operations. Things such as customer, inventory, and purchase data fall into this category. This type of data is straightforward and will generally look the same for most organizations. To know the most up to date information on something, operational data must be used. Operational data refers to all data items that are needed by a particular activity as in [16].

4.2 Proposed Approach

The proposed approach in this chapter is based on an ad hoc method applying an active help with the concept DataRecord for verification and will be able to handle loops. We suggest improving this method in order to be able to apply Rules on a loop. Additionally, an Xor-split is used by the model as a connector to feedback the exchange data flow when an error message occurs. Indeed, we propose to use the Xor-split as a decision node with a data connector at an input data I(d). Moreover, this data connection is a routing decision which can be made based on a set of data items inputted to the decision node. Then, for a function to return the data when the feedback with a read operation in the activity A2. Indeed, the error message occurres in A2. However, this proposition needs the DataRecord concept which can be changed by a Data-State. Additionally, this Data-State is initialized for each iteration and carries the latest activity and state of a dataset. The verification of data flow anomaly detection is the same, we keep an active help in an ad hoc approach. Indeed, each activity has an input data I(d) formalized as $(R_A,0,0)$, d is read in the activity A, and an output O(d′) in activity A′ as $(0,W_{A'},0)$; d′ is written in activity A′, e.g. if I(d) is accessed by activity A an output data O(d′) is produced by activity A′. Otherwise, an input I(d) must be processed for the next task and data will not change. Thus, an input data I(d) is read in the decision node and data d is chosen to be read in the activity A2 as in Fig. 3. In order to solve the problem discussed in the previous section, the Decision Node N1 is considered as a required activity by adding an input data I(d) as a decision variable. Additionally, a Boolean logic predicate (Yes, No) with a guard to specify the routing of a Decision Node N1 that (Yes = true & No = false) is added [24, p 333].

4.2.1 The Solution with a Decision Node

We began drawing the model for each task at the moment of modeling applying the proposed improvement of the approach. Therefore, if the error e exists (e ≠ Ø) the guard is "No". So, in this instance, the system must be on feedback loop involving a read data d (destination) in the activity A2. This data d has been an input I(d) in the activity N1 (decision node) that it couldn't be written or updated in A2. Otherwise, (e = Ø) the guard is "Yes", and no change in the next activity A3, the model continues processing. Consequently, the loops and Xor can only read the data; it can't delete nor create nor update data. In this case, the given data flow anomaly is only a missing data. Consequently, the missing data rule would be improved, and the DataState concept manages the state of each data in the activities which have been reinitialised in each iteration, as in Fig. 3.

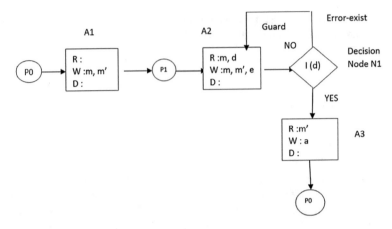

Fig. 3 The conditional node decision

4.2.2 Missing Data Rule

For each iteration, DataState is initialized to record the new stored data set and activity and the decision node N1 has a guard No = false, the data d is read in activity A2, involving the missing data.

if $d \rightarrow I(d)$ & Pred (No), then d is read in the next activity; therefore, involving missing data.

4.2.3 Message Error Rule

When an error e occurs in a task, an output is O(e), and the process cannot continue their task until the next activity due to the error e. Consequently, the loop starts, and if the guard is false (No), the system is loaded to correct the error at the next activity A2 (corrected email) in time of looping by destroying it.

4.2.4 Interpretation

We assume that error e is corrected when the system returns to activity A2 by applying Message Error Rule, and error e is destroyed. At the same time when the preceding task is connected to node N1 by an input data d, the system is to return the message error by Boolean false = No to correct the error, that involve reading data d in activity A2. Applying Missing Data Rule, the system detects an anomaly d in A2 that is a missing data. Consequently, after verification and locking the system, the modeler corrects the anomaly and chooses to write data d in activity A2 as in Data-State in Table 7.

Table 7 The new data-state with loop and decision node

Data	State	Iteration
m	$(R_{A2}, W_{A2}, 0)$	1
m'	$(0, W_{A2}, 0)$	1
e	$(0, W_{A2}, D_{A2})$	1
d	$(0, W_{A2}, 0)$	1

The decision Node has a connector input data I(d) which does not depend on the number of iterations. Consequently, our approach with a (Missing Data Rule & Message Error Rule) is valid for n iterations:

when $n > 1$, $n \in N- \{0\}$ n iterations as in Table 8

Table 8 The new Data-State with loop and Decision Node

Data	State	Iteration
m	$(R_{A2}, W_{A2}, 0)$	n
m'	$(0, W_{A2}, 0)$	n
e	$(0, W_{A2}, D_{A2})$	n
d	$(0, W_{A2}, 0)$	n

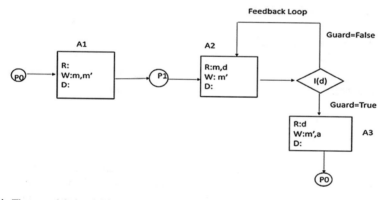

Fig. 4 The remodeled activities

4.3 Verification of Conflicting Data and Redundant Data

In the second iteration, when the errors e is corrected, the system continues the processing in the same model. There are some remodeling in the activity A2, the error e disappears, and the data d is corrected. In this mean, after passing through the decision node, the data d is still readable in activity A3. The decision node is a fragment processing, which can allow the processing to continue modeling with the data flow as in Fig. 4. In this case, we can detect the other anomalies of conflicting data and redundant data applying the rules to verify each one of the anomalies.

Table 9 The new Data-State with medication is activity A2 and A3

Data	State	Iteration
m	$(R_{A2}, W_{A1}, 0)$	2
m'	$(0, W_{A2}, W_{A3})$	2
a	$(0, W_{A3}, 0)$	n
d	$(W_{A2}, R_{A3}, 0)$	n

4.3.1 Conflicting Data Rule

Starting from the second iteration until n, $n \in N - \{0,1\}$, for a given data operation d of data flow.

(While error e = 0) do

If data d is writing in the activity A also in activity A' and in the rest of the next activities without an update. Also, these data aren't read in the following activities. That involves => conflicting data.

4.3.2 Redundant Data Rule

When the iteration = n, $n \in N - \{0,1\}$,

(while guard = true) do

If the data d is written in the last activity, then data d is recorded in data-state in write, and never had read, that involves => Redundant data.

4.3.3 Interpretation of Two Rules

Applying the approach, our assessment in this case, at each fragment of the process the system is locked to correct the detected anomalies. Consequently, it's confirmed that data m' is an anomaly conflicting and data a is an anomaly redundant as shown in Table 9.

4.4 Results of Rules Implementation

After theoretical interpretation of the verification by applying missing data Rule of our model for processing; the rules are to be implemented and verified in a real tool. Therefore, we propose to use the KIE (Knowledge Is Everything), that is an umbrella project introduced to bring our related technologies together under one roof. It also acts as the shared core between our projects. Indeed, KIE contains the following different but related projects offering a complete portfolio to implement the Rules. "*The modularization of the knowledge base which helps in managing rules and improves the efficiency of rule-based system execution. In CLIPS each module has its own pattern-matching network for its rules and its own agenda. When a run command is given, the agenda of the module, which is the current focus, is executed. Before a rule is executed, the current module is changed to the module in which the executing rule is defined (the current focus)"* [26].

Drools is a business rule management system with a forward-chaining and backward-chaining inference-based rules engine, allowing fast and reliable evaluation of business rules and complex event processing. *"The Drools platform introduced RuleFlow tool. It is a workflow and process engine that allows for the advanced integration of processes and rules. It provides a graphical interface for processes and rules modeling. Drools have a built-in functionality to define the structure of the rule base which can determine the order of the rules evaluation and execution."* [26].

jBPM is a flexible Business Process Management suite allowing us to model our business goals by describing the steps that need to be executed to achieve those goals. Indeed, *"jBPM is an open-source business process management suite, embedded in the KIE group, which executes repeatable workflows. This solution is Java EE based and it runs as a Java EE application. The system supports multi-user collaboration, using groups of users, but its configuration is rather complex for users without technical skills"* [27].

Drools Workbench is a full featured web application for the visual composition of custom business rules and processes.

Some users of the tool jBPM Drools faced some problems when installing it in their computers, so we choose the version jbpm-server-7.14.0. Consequently, in order to finish this installation, it should be connected with the address as in Fig. 4.

After the connection to KIE, it must be identified by admin or the others login found in the readme notice, to access the menu.

Identification with Admin in interface of KIE:

Login: admin

Password: ....

Finally, the menu KIE Workbench is open. It has four submenus: Design, Deploy, Manage and Task. Indeed, to start work, the user must choose one of these submenus. In this case, we select the menu Design for drawing our process model. Additionally, in the menu Design, we select the project and choose the Authoring Perspective to create a new project. Then, we enter our project details. After a project has been created, we need to define Types to be used by our rules. Select "Data Object" from the "New Item" menu. We start with the first model with xor-split.

To define the missing data rule, we select "DRL file" (for example) from the "New Item" menu. We then enter a definition for missing data rule. Then we implement our jBPM Drools associated with our work to execute the Rules in the process model. The results are shown in Fig. 5.

Therefore, when we choose the spaces of Data_Flow_Anomalies of Missing Data Rule.rdrl -Guided Rules

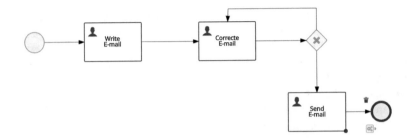

Fig. 5 The first figure without decision node

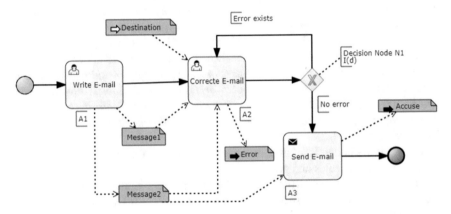

Fig. 6 The process model with the decision node

The algorithm of Missing data Rule from Drools is given bellow:

Import java.lang.Number ;

Rule "Missing Data Rule "

    Dialect "mvel "

      when
               $i : iteration()
               $p :Pred(iteration ==$i)
      then
               insert (new MissingDataError())
    end

Finally, after entering the rule in our KIE workbench, we have the last view of the model with a decision node as in Fig. 6.

In the meantime, we have verified only the missing data rule by the system KIE. However, the conflicting data rule and the redundant rule need also to be verified.

Thus, we propose to verify the rules by another tool which we will use in the future work. In the following subsections we discussed some definition of uppaal.

4.4.1 Implementation with Model Checking

we verify the conflicting data and redundant data rules using a tool Uppaal of model checking. For each instance, the active help is applied in the model of Fig. 4; at the moment the anomalies are detected the system is locked and data-state records the last state for data and the activity. In this time when the modeler chooses that errors to be corrected as proved in Fig. 4, and the system continues for modeling that we validate with the simulator in Uppaal. A computation tree logic CTL* in a temporal logic using subset LTL is to be investigated, as well as our example on which the model can validate as in [15, 28, 29].

To model the function of the two remaining rules, Uppaal is an environment that is integrated which permit to simulate and analyze the verification of models in particular states and in transitions between states as in [30].

In this way, we use in Uppaal a simulation of two template Messenger and SendMess. This template uses for one of them the two instances messag1 and message2 for Messenger and for SendMess there are Sendmessg1 and Sendmessg2 as show in Table 10.

In this chapter, we use the absolute basics of Uppaal to validate the model and to verify the conflicting data rule and redundant data rule. Consequently, we cited the states and the transitions used in the system to construct the automates used in the tool Uppaal (Table 11).

Table 10 Description of templates and instances used in Uppaal

Templates	Instances
Messenger	Mcssage1
	Message2
SendMess	SendMessg1
	SendMessg2

Table 11 Description of states and transitions used in Uppaal

States (places)	Transitions
start	Open_emai
First_message	Write_message1
Seconde_message	Write_message2
Correct_error_e	Correct_email
Decision_node	Feedback_correction:error_e
Write_destination	Send_email
end	Close_email

In this way, the Uppaal graphical consists of the interface of main parts that are accessible via three tabs in the main window: the system editor to constructs models, the simulator in order to input the system being simulated, finally the verifier in which the cited system rules are verified.

In this subsection, we discuss the system editor, also we present the simulator of the two templates messenger and SendMess with their instances and show message1 with SendMessg2 and the message2 with SendMessg1 as in the Figs. 7, 8, 9 and 10.

Message1 = Messenger ();
Message2 = Messenger ();
SendMessg1 = SendMess ();
SendMessg2 = SendMessg();

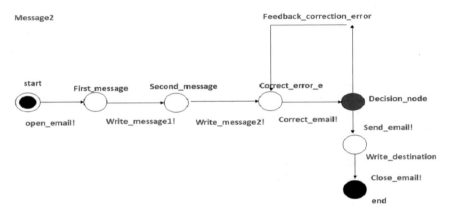

Fig. 7 Simulation of template messenger

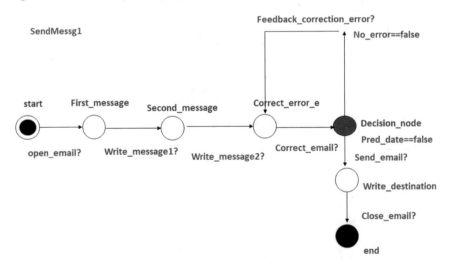

Fig. 8 Simulation of template SendMessg

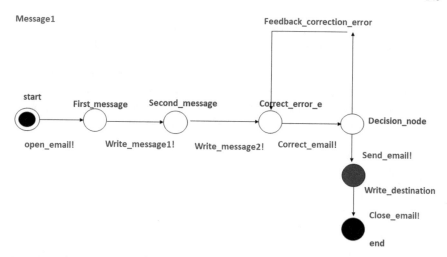

Fig. 9 Simulation of template messenger

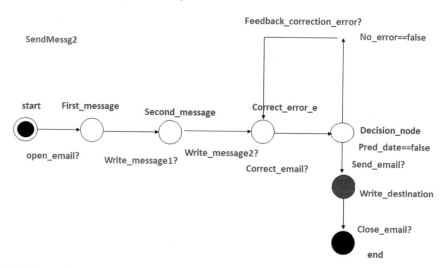

Fig. 10 Simulation of template SendMessg

Upon starting the simulator, the system chooses witch instances are synchronised to gether. After the simulator is started, we began to select the transition for two templates coloured red until the system is deadlock as in Figs. 9 and 10.

"Binary synchronisation channels are declared as chan c; c is a transition. An edge labelled with c! synchronises with another labelled c?. A synchronisation pair is chosen non-deterministically if several combinations are enabled as in [31]".

The idea is to model a system using timed automata, simulate it and then verify the properties on it. Timed automata are limited states automated with time (clocks) as in [31].

The result of this simulation is to validate our automates. The time where we click **next** the simulator continues to respect the model in Fig. 4 using the two templates chosen in Uppaal as in [30–33].

In this way, for each click of next, another chart mentioned below to construct at the same time as the simulation defined the scenarios of the passage the places (states) progressing the one transition to another; respecting the routing of the process by the running simulator until the end of the model as shown in Fig. 11.

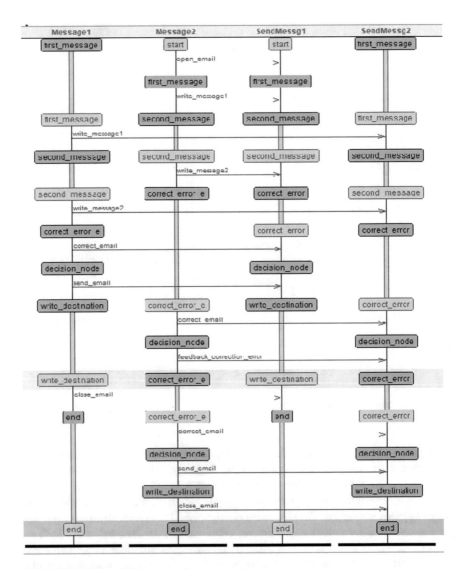

Fig. 11 The simulation is shown by the sequence diagram

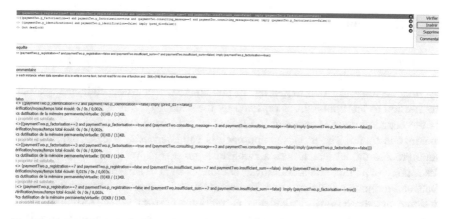

Fig. 12 The verification of rules

In this implementation we began to draw the templates and to obtain an insight of validation, the template simulator played this role, but our object is not atteint yet. Therefore, the part of verifier on the tool is important for approved the existing anomalies when modeling. The requests predicates in this part check the deadlock running of each simulation and the conflicting data rule and the redundant rule.

In this case, we use the properties for Uppaal [30–32] to verify these rules as in Fig. 12.

- $E <> p$: there exists a path where p eventually holds.

The verification of model is:

$$E <> notdeadlock$$

Conflicting data Rule:

$$E <> \left(\begin{aligned} (SendMessg2.correct\_error == 2 and SendMessg2.write\_destination == false) \\ imply(pred\_data == false) \end{aligned} \right)$$

Redundant data Rule:

$$E <> \left(\begin{aligned} (SendMessg2.correct\_error == 8 and SendMessg2.write\_destination == false) \\ imply no\_error == true \end{aligned} \right)$$

When the request is inputted in the verifier and we click in bouton verifier, the system checks the rule and writes the propriety is satisfied. In this sense, we tried to verify the two rule, confliction data and redundant data as chow in Fig. 12.

5   Conclusion

In this chapter, we developed an active help with a Data-Record concept to detect data flow anomalies in loop modeling. We proposed to improve the active help approach by suitable rules for loop modeling. In this context, a decision node, using a data connection as an input data, is replaced by the connector Xor-split. The input data of the decision node is returned to the last activity by a feedback when the error message is found. The proposed approach is validated using a deterministic finite state process model which uses a Boolean logic predicate (Yes or No) to specify the routing of an input data. Moreover, anomalies such as Missing Data, Conflicting Data and redundant Data are used. The verification is triggered when an anomaly is detected, and the system is locked until correction is performed. The results show that missing data anomalies were efficiently handled. Their implementation was verified by Kie workbench and jBPM. Therefore, the Conflicting data and redundant data anomalies are verified by another tool for model checking that is Uppall.

References

1. Basu, A., Blanning, R.W.: A formal approach to workflow analysis. Inf. Syst. Res. https://pubsonline.informs.org (2000)
2. Burattin, A.: Introduction to business processes, BPM, and BPM systems. Process Mining Techniques in Business Environments, pp. 11–21. Springer, Berlin (2015)
3. Bruno, G.: Data flow and human tasks in business process models. Procedia Comput. Sci. **64**, 379–386 (2015). Elsevier
4. Davenport, T.H.: Reengineering Work Through Information Technology. Harvard Business School Press, Boston. https://is.ieis.tue.nl (1993)
5. Rausch, G.K.P.L.S., Retschitzegger, S.W.: Workflow management based on objects, rules, and roles. In: IEEE Bulletin of the Technical Committee on Data Engineering (2016)
6. Hai-yan, X., Yan, W.: Workflow model based on stochastic Petri nets and performance evaluation. In: IEEE International Symposium on IT in Medicine Education, 2009, ITIME'09, vol. 1, pp. 245–249 (2009)
7. Gabriel, A., Camargo, M., Monticolo, D: Process modelling for a creative problem-solving support system. In: Proceedings of the 2014 International Conference on Innovative Design and Manufacturing (ICIDM) https://ieeexplore.ieee.org (2014)
8. Guo, X., Sun, S., Vogel, D.: A dataflow perspective for business process integration. ACM Trans. Manag. Inf. Syst. **5**(4), (2015). Article 22
9. Cong, L.I.U., Zeng, Q., Hua, D.: Formulating the data-flow modeling and verification for workflow: a petri-net based approach. Int. J. Sci. Eng. Appl. **3**, 107–112 https://www.ijsea.com (2014)
10. Hua, R., Fu, Y., Yu, J.Z., Liu, C.: Petri Net-based modeling and verification of automatic train speed control system. Appl. Mech. Mater. **571**, 395–399 (2014). Trans Tech Publications
11. Xiang, D., Liu, G., Yan, C., Jiang, C.: Detecting data-flow errors based on Petri nets with data operations. IEEE/CAA J. Autom. Sin. **5**(1), 251–260 https://ieeexplore.ieee.org (2018)
12. Sun, S.X., Zhao, J.L., Nunamaker, J.F., Sun, S.X., Zhao, J.L., Nunamaker, J.F.: Formulating the data-flow perspective for business process management. Inf. Syst. **17**(4), 374–391 https://pubsonline.informs.org (2006)
13. Kabbaj, M.I., Betari, A., Bakkoury, Z., Rharbi, A.:, Towards an active help on detecting data flow errors in business process models. IJCSA **12**(1), 16–25https://www.researchgate.net (2015)

14. Landmark, L., Hauge, M., Kure, O.: Routing loops in mobile heterogeneous Ad Hoc networks. In: IEEE Military Communications Conference, MILCOM 2013, https://ieeexplore.ieee.org (2013)
15. Trcka,N., van der Aalst, W., Sidorova, N.: Analyzing control-flow and data-flow in workflow processes in a unified. Computer science report, https://research.tue.nl (2008)
16. Sadiq, S., Orlowska, M., Sadiq, W., Foulger C.: Data flow and validation in workflow modelling. In: Proceedings of the 15th Australasian Database Conference, https://dl.acm.org (2004)
17. Wang, F.J., Hsu, C.L., Hsu, H.J.: Analyzing inaccurate artifact usages in a workflow schema. In: Computer Software and Conference, https://ieeexplore.ieee.org (2006)
18. Meda, H.S., Sen, A.K., Bagchi, A.: Detecting data flow errors in workflows: a systematic graph traversal approach, https://papers.ssrn.com (2007)
19. Trčka, N., Van der Aalst, W.M.P., Sidorova, N.: Data-flow anti-patterns: Discovering data-flow errors in workflows. In: International Conference on Advanced Information Systems Engineering, Springer, Berlin (2009)
20. Weske, M.: Business process management architectures. Business Process Management. Springer, Berlin (2012)
21. Rgibi, A.E., Yao, S.Z., Xu, J.J.: Dataflow errors detection in business process model. Applied Mechanics and Materials. Trans Tech Publications (2012)
22. Rgibi, A.E.S.: Data flow issues and BPMN mapping to Petri Net: Road map. In: Proceedings of the Industrial Engineering, https://www.taylorfrancis.com (2015)
23. Dolean, C.C., Petrusel, R.: Data-flow modeling: a survey of issues and approaches. Informatica Economica, https://www.researchgate.net (2012)
24. He, Y., Liu, G., Xiang, D., Sun, J., Yan, C., Jiang, C.: Verifying the correctness of workflow systems based on workflow net with data constraints. IEEE Access, https://ieeexplore.ieee.org (2018)
25. Sun, S.X., Zhao, J.L.: Formal workflow design analytics using data flow modeling. Decision Support Systems, Elsevier (2013)
26. Nalepa, G.J.: Integrating business process models with rules. Modeling with Rules Using Semantic Knowledge Engineering, Springer, Berlin (2018)
27. Almeida, J., Ribeiro, R., Oliveira, J.L.: A modular workflow management framework. In: Proceedings of the 11th Innovations in Software Engineering, https://www.researchgate.net (2018)
28. Clarke, E.M., Grumberg, O., Peled, D.: Model checking, https://books.google.com (1999)
29. Emerson, E.A., Halpern J.Y.: On branching versus linear time temporal logic. J. ACM (JACM), https://dl.acm.org (1986)
30. Vaandrager F.: A First Introduction to uppaal. Deliverable no: D5. 12 Title of Deliverable: Industrial Handbook (2011)
31. David, A., Amnel, T., Stigge, M., Ekberg P.: Abgerufen am, UPPAAL 4.0: Small Tutorial. A tutorial on Uppaal 4.0 (2011)
32. Mendoza Morales, L.E.: Business process verification: the application of model checking and timed automata. CLEI Electron. J. http://www.scielo.edu.uy (2011)
33. Wernhard, C.: The Boolean solution problem from the perspective of predicate logic. In: International Symposium on Frontiers of Combining, Springer, Berlin (2017)

An Adaptive Vehicular Relay and Gateway Selection Scheme for Connecting VANETs to Internet via 4G LTE Cellular Network

Driss Abada, Abdellah Massaq, Abdellah Boulouz and Mohamed Ben Salah

Abstract In this paper, we propose an efficient relay and mobile gateway vehicle selection in heterogeneous VANETs and 4G LTE cellular networks. The proposed scheme is based on timer-based contention strategy to elect some potential vehicles equipped with both 4G LTE and IEEE 802.11p-based-VANETs interfaces among several other neighboring to act as vehicular gateways in the network, for providing Internet access to source vehicles moving in the road. Therefore, if one source vehicle does not have any gateway in its coverage, the proposed approach integrates an efficient relay selection scheme to select some vehicles equipped with IEEE 802.11p interface to act as relays in order to reach transmission range of mobile gateways. Stability, mobility and received signal strength are the main related metrics considered in the relay and mobile gateway vehicle selection.

Keywords VANET · 4G LTE · Routing · Vehicular gateway selection · Vehicular relay selection · Heterogeneous networks

D. Abada · A. Boulouz · M. Ben Salah
Laboratory LABSIV, Faculty of Sciences, Ibn Zohr University, Agadir, Morocco
e-mail: drissmisr@gmail.com

A. Boulouz
e-mail: abdellah.boulouz@gmail.com

M. Ben Salah
e-mail:m_ben_salah@yahoo.fr

A. Massaq (✉)
Laboratory OSCARS, National School of Applied Sciences,
Cadi Ayyad University, Marrakesh, Morocco
e-mail: a.massaq@uiz.ac.ma

© Springer Nature Switzerland AG 2020
M. Elhoseny and A. E. Hassanien (eds.), *Emerging Technologies for Connected Internet of Vehicles and Intelligent Transportation System Networks*, Studies in Systems, Decision and Control 242, https://doi.org/10.1007/978-3-030-22773-9_10

1   Introduction

Internet access in Vehicular Ad hoc Networks (VANETs) [1] has recently become an increasingly popular research topic in the area of wireless networking as well as the automobile industries, given the more new important safety and entertainment applications that can be enabled for improving Intelligent Transportation System (ITS) [2].

IEEE 802.11p [3, 4] is a protocol designed to make reliable communication between neighboring vehicles equipped with On-Board Unit (OBU), i.e Vehicle to Vehicle (V2V) or between vehicles and nearby Road Side Unit (RSU), i.e. Vehicle to Infrastructure (V2I) communications. This standard has introduced the modifications at the physical and MAC layers of IEEE 802.11. The objective is reducing the effects of Doppler shift, caused by the fast mobility of the vehicles, and eliminating the inter-symbol interference, due to multipath fading. It uses the DSRC channel, which is designed specifically for short range and time-sensitive applications to support the high mobility of the nodes. Long Term Evolution (LTE) [5] is a 4G wireless broadband technology developed by the Third Generation Partnership Project (3GPP). LTE cellular network is a well-designed system that offers high peak data rates that could reach 1Gb/s and presents a low latency of about 5ms and large coverage of communication of around 8–10 km.

In the high mobility scenarios, using only IEEE 802.11p as Internet access technology, the network may suffer from a lot of issues such as intermittent connectivity, fast topology changes, reliability requirement dissatisfaction and frequent link failures which will affect negatively the network performance and quality of Internet connection. To overcome these problems, the researchers thought to integrate both IEEE 802.11p and LTE cellular network technologies in a heterogeneous network. In this case, the high data rates of IEEE 802.11p-based VANETs and the wide coverage area of 3GPP networks (e.g., LTE) are coupled. Integration of VANETs and 3GPP networks is addressed in many research works [6–8]. Most of these works used clustering technic to make efficient communication and routing in VANETs. However, the clustering in VANET has drawbacks such that fast topology change, intermittently connected network and various communication environments which make the clustering scheme costly in terms of explicit control messages required for formation and maintaining clusters in the network.

In this paper, we propose an integrated VANETs-LTE heterogeneous network architecture. The focus of this paper is proposing a distributed mobile gateway selection approach that selects the vehicles equipped with IEEE 802.11p and LTE interfaces as vehicular gateways to connect ordinary vehicles with the 4G LTE network. Therefore, the route selection to the vehicles designated as vehicular gateways in the network and discovery is performed through an efficient multi metrics relay selection scheme, in case that there is no vehicular gateway within the transmission range of source vehicle. Stability of links, receiver signal strength (RSS) of vehicles and mobility speed are all taken into consideration when selecting vehicle relays and gateways using Contention-Based Forwarding (CBF) strategy [9].

Accurately, the public transport taxis or buses can be used to act as mobile gateways on the road. Many feasibility studies [10–12] of using public transport vehicles as mobile gateways in the road to provide Internet access in VANETs are performed.

The remainder of the paper is organized as follows: Some related works are cited in Sect. 2. In Sect. 3, we explain and illustrate the adopted system model. The suggested mechanism of relay and gateway selection is detailed in Sect. 4. Performance evaluation of the proposed scheme is given to Sect. 5. Finally, we give a conclusion in Sect. 6.

2 Related Networks

In the literature, several works have proposed to make communication within VANETs network more reliable, by integrating 3GPP and VANETs wireless technologies.

The enhanced hybrid wireless mesh protocol (E-HWMP) suggested in [13] is used to improve mobile gateway selection mechanism in order to perform an efficient multi-hop routing via VANETs to reach the LTE base station. LTE received signal strength, available route capacity and stability are the three metrics used for selecting the optimum gateway vehicle. In the objective of making efficient multi-hop vehicular communication, the authors have based on the cluster-based approach. Moreover, the path to the vehicular gateway discovery is hybrid. These improvements are incorporated in the protocol for providing Internet access in VANETs with reducing the overhead and delay and increasing packet delivery.

The focus of the work proposed in [14] is suggesting a new UMTS-VANETs integrated network architecture. The proposed approach consists of group vehicles dynamically based on multiple metrics and then select the most appropriate of them to play the role of gateways for connecting VANETs to the Internet via UMTS base stations. The gateway selection technique aims to prevent as possible overloading UMTS base station, while the clustering-based approach enables to efficiently manage inter-vehicular communication and reduce the redundancy effect of unnecessary messages. However, selecting the cluster head as the vehicle closest to the center of the cluster may not be the best solution because, depending on the speed of the vehicle, it can quickly move away from the center. In addition, the clustering formation and cluster head selection mechanisms in this work are complex and may require more signaling traffic and more time than those for data traffic exchange.

In the research work [15], the authors proposed a routing protocol for connecting vehicular networks to the Internet through a stationary gateways which are installed along the road. The proposed protocol used two important metrics, namely, link expiration time (LET) and route expiration time (RET), to select the most stable route between source vehicles and gateways. By using these metrics, the protocol aims, proactively to propagate the advertisement messages through multi-hops without flooding the network, does seamless hand-overs and selects the most stable routes to these static units. However, protocols based on fixed gateway called roadside unit (RSU) are not more efficient to connect VANETs to the Internet, because, due to the drawbacks of 802.11p cited in the previous section, it suffers from continues intermittent connection, frequent handover and need to install many RSUs along the road to provide Internet connection. In the present work, we will replace this static RSUs with public transport taxis or buses which are used as mobile gateways [10]. In the contrast, these RSUs can be very useful for safety and traffic related applications of VANETs, such as jam detection, weather forecast, fraud detection, etc.

3   System Model

The system model of the envisioned VANET-LTE integrated network architecture is based on a bidirectional multi-lane highway road scenario as illustrated in Fig. 1. The architecture comprises vehicles and LTE Evolved Node B (eNodeB) base station. In this architecture, we distinguished four types of vehicles: (1) ordinary vehicles (OV)

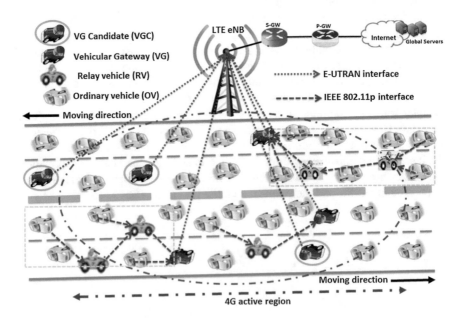

Fig. 1 Envisioned VANET-LTE integrated network architecture

that are equipped with OBU that contains only IEEE 802.11p wireless interface used to communicate with its neighboring vehicles (2) relay vehicles (RV) are the ordinary vehicles that are selected to act as relays between sources and gateways (3) vehicular gateways candidates (VGC) are the vehicles that are equipped with both IEEE 802.11p and LTE capabilities which allow them to communicate in VANETs using IEEE 802.11p interface and with the eNodeB using EUTRAN (Evolved Universal Terrestrial Radio Access Network) interface and (4) vehicular gateway (VG) are the vehicular gateway candidates that are selected to act as Internet mobile gateway in the network. Moreover, vehicles are assumed to be equipped with a GPS system, allowing them to obtain their locations, speeds, and directions.

The main goal of this architecture is among several neighboring designated vehicular gateways, we aim to select in a distributed manner one potential vehicle to act as a vehicular gateway in the current coverage. As a consequence, a minimum number of gateways, per time instance, is selected to connect ordinary vehicles with the LTE network. Metrics that reflect VANETs characteristics such as link connectivity duration, received signal strength, mobility speed and distance rate are all taken into account in the selection. These vehicles will play a proxy role in the network to connect vehicular sources to the Internet via LTE infrastructure.

4 Proposed Relay and Gateway Selection Scheme

4.1 Metrics of Gateway Selection

4.1.1 Mobility Speed of VGC

The speed of mobility metric is used as one of the considerable metrics of selection. The vehicular gateway candidate which moves with low speed are more appropriate vehicles desired to be selected as gateway, because it may increase network connectivity and decrease routing overhead due to minimizing the frequent handover and the re-selection of new gateway. For computing the average speed of mobility metric, we do a speed sampling during the time, memorize and employ the data to determine speed values. For each vehicle we extract n samples of speed $(v_0, v_1, \ldots, v_{n-1})$ every τ seconds at instant $t_0, t_1, \ldots, t_{n-1}$. As time proceeds, we drag the sampling window to have the information recently stored. The mobility speed metric is calculated as follow:

$$M_1 = 1 - \frac{\overline{V}_{VGC}}{V_{max}} \tag{1}$$

where V_{max} is the maximum speed in the network and $\overline{V}_{VGC}$ is the mean of sampled speed values, with,

$$\overline{V}_{VGC} = \frac{\sum_{i=0}^{n-1} v_i}{n} \tag{2}$$

4.1.2 Received Signal Strength (RSS) of VGC

Received Signal Strength (RSS) of VGC is the eNodeB base station received power level detected by the VGC. This metric is used to select the vehicular gateways candidates that has best RSS as vehicular gateways in the network. Similar to the mobility speed, we take n samples of RSS ($RSS_0, RSS_1, ..., RSS_{n-1}$), at the same instances $t_0, t_1, ..., t_{n-1}$. The metric M_2 is calculated using the following formula:

$$M_2 = 1 - \frac{RSS_{th}}{\overline{RSS}} \tag{3}$$

where RSS_{th} is the received signal strength threshold and $\overline{RSS}$ is the average RSS. Each VGC can calculate its current RSS at instant t from LTE BST using equation proposed in [14].

$$RSS_t = RSS_{t-1} \pm \left(1 - e^{\frac{|v_t - v_{t-1}|}{a}}\right) \tag{4}$$

- RSS_t and RSS_{t-1} denote the values of the LTE signal strength received at time instances t and $t - 1$, respectively,
- v_t and v_{t-1} denote the values of the mobility speed of the vehicles at time instances t and $t - 1$ such that $0 < v_t, v_{t-1} \le V_{max}$, where V_{max} is the maximum speed of the vehicle,
- a is a constant that defines the rate of variation of the LTE signal strength for a unit increase or decrease in the mobility speed, in a particular movement direction, relative to the position of the LTE BST.
- (+) denotes that the vehicle is moving toward LTE-BST, whereas (−) denotes that the vehicle is moving away from LTE-BST.

4.2 Vehicular Gateway Selection Algorithm

In this work, we have based on the timer-based strategy to elect in a distributive manner potential vehicular gateways in the network. The strategy used here is called CBF (Contention-Based-Forwarding). We suppose that the clocks of all vehicles are synchronized. At the end of periodic interval time, the neighboring vehicular gateways candidates do not broadcast the advertisement (ADV) message instantly, but defer their broadcasting by a given timer and come into a contention phase. The first VGC which waiting timer terminates, that instant broadcasts the message and its status becomes VG and any other VGC overhearing that transmission and its movement direction is the same to that of the sender, cancels its timer and does not broadcast. As a consequence, a minimum number of VGCs will be selected as VGs in the network.

The following expression represents the weighting function denoted W_{VGC} that combines both mobility speed and RSS metrics using weithed mean with factor $\alpha \in [0, 1]$:

$$W_{VGC} = \alpha.M_1 + (1 - \alpha).M_2 \tag{5}$$

Basing on the computed value of W_{VGC}, each vehicle sets its timer $t(.)$ using the following formula:

$$t(W_{VGC}) = T_{max} \times (1 - W_{VGC}) \tag{6}$$

where T_{max} is a maximum time forwarding. Note that more W_{VGC} takes high values, more than the waiting time is decreasing. Thus the vehicle will have a high chance to be the first broadcaster advertisement message and becomes a potential gateway. The pseudo code of the proposed distributed vehicular gateway selection scheme is presented as in Algorithm 1.

Algorithm 1: Algorithm of Vehicular Gateway Selection

---

1 **begin**
2 │ All vehicular gateways candidates set their status to VGC and ordinary vehicles to OV.
3 │ **foreach *End of Broadcast interval time of ADV message* do**
4 │ │ **if** *RSS of VGC* $\geq$ *RSS$_{th}$* **then**
5 │ │ │ Calculate the both metrics M_1 and M_2 ;
6 │ │ │ Compute a waiting time $t(.)$;
7 │ │ │ Run timer;
8 │ │ **end**
9 │ │ **if** *the waiting time is expired* **then**
10 │ │ │ status =VG;
11 │ │ │ The vehicle broadcasts ADV message including speed,position and direction;
12 │ │ **end**
13 │ │ **if** *the ADV message is received && the vehicles move in the same direction* **then**
14 │ │ │ **if** *status==VG* **then**
15 │ │ │ │ Drop message;
16 │ │ │ **else**
17 │ │ │ │ **if** *status==VGC* **then**
18 │ │ │ │ │ Cancel the waiting timer and drop message;
19 │ │ │ │ │ no change in status;
20 │ │ │ │ **else**
21 │ │ │ │ │ the vehicle executes algorithm 2;
22 │ │ │ │ **end**
23 │ │ │ **end**
24 │ │ **end**
25 │ **end**
26 **end**

4.3 Metrics of Relay Selection

4.3.1 Stability Metric

This metric is used to select most stable routes between vehicular sources and gateways. To reflect the stability of the links, we take advantage to Link Connectivity Duration (LCD) which represents the time duration in which two vehicles at each end of the link are within each others transmission range in the network. The stability metric can be calculated using the following formula that is proposed in [15]:

$$M'_1 = 1 - e^{-2\frac{LCD}{RCD}} \tag{7}$$

where LCD is the duration that two vehicles i and j remain in connection. If we suppose that all vehicles have the same communication range r, the LCD is computed using this formula, inspired from [15]:

$$LCD_{ij} = \frac{\sqrt{(a_{ij}^2 + c_{ij}^2)r^2 - (a_{ij}d_{ij} - b_{ij}c_{ij})^2} - (a_{ij}b_{ij} + c_{ij}d_{ij})}{a_{ij}^2 + c_{ij}^2} \tag{8}$$

where,

- $a_{ij} = v_i \cos\theta_i - v_j \cos\theta_j$
- $b_{ij} = x_i - x_j$
- $c_{ij} = v_i \sin\theta_i - v_j \sin\theta_j$
- $d_{ij} = y_i - y_j$.
- (x_i, y_i) and (x_j, y_j) are the Cartesian coordinates of locations of two vehicles i and j respectively.
- v_i and v_j are the velocities of vehicles i and j.
- θ_i, θ_j are the direction angles of vehicles i and j.

If the vehicles i and j are not adjacent, the LCD is the minimum of LCD_k ($1 < k < h$), where h is the number of hops between the two vehicles. We call this metric the Route Connectivity Duration (RCD).

4.3.2 Inter-Vehicular Distance Metric

This metric is used to select the furthest next-hop vehicle by choosing among neighbor nodes that are within a predefined maximum transmission range. Inter-vehicular Distance metric will be combined with LCD in order to select shortest path in term of hops with maximum lifetime route to the vehicular sources. We can calculate inter-vehicular distance metric between the source vehicle i and its neighbors j as follows:

$$M'_2 = \frac{dist_{ij}}{r} \tag{9}$$

with, $dist_{ij} = \sqrt{(x_i - x_j)^2 + (y_i - y_j)^2}$.

4.4 Vehicular Relay Selection Algorithm

Similar to the gateway selection scheme, the relay selection scheme based on the CBF strategy. Algorithm 2 shows the pseudo code that is used to make a decision of the appropriate OVs that might be selected as relays in the network, to allow source vehicles that did not have any vehicular gateway within their coverage. Upon reception of ADV message, before entering in contention phase, each neighboring OV computes its weighting function W_{OV} and set its timer $t(.)$ using following formulas.

$$W_{OV} = \beta.M_1' + (1 - \beta).M_2' \tag{10}$$

where, $\beta \in [0, 1]$ is the weighting factor.

$$t(W_{OV}) = T_{max} \times (1 - W_{OV}) \tag{11}$$

where T_{max} is a maximum waiting time.

Algorithm 2: Vehicular Relay Selection Algorithm

---

1 **begin**
2   **On receiving ADV message from VG or RELAY vehicle;**
3   **if** *The received ADV message is new && the vehicles move in the same direction* **then**
4   Compute M_1' and M_2' ;
5   Calculate W_{OV} ;
6   Compute a waiting time $t(.)$;
7   Run timer;
8   **else**
9   Cancel the waiting timer and drop both messages;
10   **end**
11   **On the waiting time expire ;**
12   Update the received message with new information ;
13   The vehicle re-broadcasts updated ADV message;
14   status=RV;
15 **end**

---

5 Routing Path Discovery to the Mobile Gateway

The proposed vehicular gateway selection scheme can be implemented on top of any VANET routing protocol. We consider the usage of the AODV+ [20] as it copes efficiently with the highly dynamic nature of VANETs. The routing path establishment is performed using the proposed relay selection scheme and maintenance procedure of routing is similar to routing protocol AODV+. In the routing path proactive discovery procedure, each vehicular gateway broadcasts periodically an ADV message which is propagated through intermediates nodes in its restricted proactive zone (number of hops ≤3). In this paper, we have improved AODV+ to select optimal routes by considering the maximum value of Link Connectivity Duration (LCD) and inter-vehicular distance rate that reflect the number of hops.

When a vehicle in the proactive broadcast zone receives the ADV message, and it does not have any entry with the address of the gateway from which the message was sent, it simply adds an entry to the routing table. If there is an entry corresponding the mobile gateway address, and the sequence number of the received message is greater than the entry in the routing table, the entry will be updated with new information included in the message. However, if a message has the same sequence number as the entry in the routing table, the remaining lifetime of entry will be compared with the RCD carried by the message. If the first is less than the second, the vehicle updates parameters in its routing table as well as in the ADV message. In case the tie, the update of entry will be performed if the message is coming with the minimum number of hops. Routes are removed from the routing table when their lifetime expires.

6 Performance Evaluation

The efficiency and the performance of the proposed approach are evaluated by exploiting the open source simulators MOVE [23], SUMO [24], and NS2 [25]. The SUMO (Simulation of Urban Mobility) is used to generate our mobility model, NS2 (Network Simulator 2) is used to implement our scheme and then measure its network performance, while MOVE (MObility model generator for Vehicular networks) is a java application that provides us a graphical interface facilitating the interaction between SUMO and NS2. The NS2 version used in this work is ns-2.33. Simulation of LTE Access and Core Networks, and its integration with IEEE 802.11p network interface is carried out patching LTE patch with ns-2.33 and using 802_11Ext and WirelessPhyExt modules which are integrated already in NS2. The Tables 1 and 2 provide a summary of all simulation environment parameters.

We give in the Table 3 some adequate values of the used parameters in our routing protocol. These values are obtained based on a lot of simulations and experiences.

For comparing the network performance of the proposed approach, we distinguished two schemes: M-AODV+-VRGSA that represents the modified version of

Table 1 Mobility features

Parameter	Value
Mobility model	Highway
Highway length	8 km
Number of lanes	2 for each direction
Maximum speed	10, 20, 30, 40, 50 m/s
Number of vehicles	200 and 300
Number of dual interfaces vehicles	Varying
Simulation time	500 s

Table 2 Integrated network parameters

Parameter	Value
Channel	Channel/WirelessChannel
Propagation model	Propagation/Nagakami (m = 3)
Network interface	Phy/WirelessPhyExt
MAC	Mac/802_11Ext
Interface queue	Queue/DropTail/PriQueue
Antenna type	Antenna/OmniAntenna
Data rate	6 Mbps
Interface queue	20
IEEE 802.11p transmission range R	300 m
Routing protocols	M-AODV+ with VRGSA and VRSA
Traffic type	CBR
Packet sending interval	0.2 s
Packet size	512 bytes
LTE	LTE patch of ns2.33
LTE RSS threshold	−94 dBM

Table 3 Simulation parameter setting

Metric	n	τ	α	β	T_{max}
Value	4	4 s	0.6	0.5	3.75 ms

AODV+ which integrates both vehicular relay and gateway selection algorithms and M-AODV+-VRSA that represents the modified version of AODV+ using only vehicular relay selection algorithm.

Packets Delivery Ratio, End to End packet Delay and Normalized Routing Overhead are the three metrics used to measure the performance of our scheme.

- The Packets Delivery Ratio (PDR) calculates the ratio between the number of data packets received by the destination vehicle and the number of packets sent by the source vehicle. It is calculated for each pair of Source-Destination vehicle and then averaged. The final equation of the Packets Delivery Ratio calculation is as follows:

$$PDR = \frac{1}{N} \sum_{i=1}^{N} \frac{R_i}{S_i} \qquad (12)$$

where N is the number of source-destination pairs communicated during the simulation time. S_i and R_i are the total number of data packet sent and received by source-destination pair i.

- The End to End packet Delay (E2ED) is the average time required data packet to reach the destination. it is obtained by counting the delays of messages received from end to end.

$$E2ED = \frac{1}{K} \sum_{i=1}^{K} Tr_i - Ts_i \qquad (13)$$

where K is the total number of received data packets, Tr_i and Ts_i are respectively the reception and transmission times of the current packet i.

– The Normalized Routing Overhead (NRO) is a ratio that evaluates the bandwidth used by data packets that have reached their destinations. Indeed, it is the ratio of the number of transmitted control packets to the number of data packets received:

$$NRO = \frac{\sum_{i=1}^{N} C_i}{\sum_{i=1}^{N} C_i + \sum_{i=1}^{N} D_i} \qquad (14)$$

were i is ID of the packet, C_i is the control packet i, D_i is the received data packet i.

6.1   Varying Number of Vehicular Gateway Candidates

First, we compare the performance of the routing protocols by changing the number of dual interfaces vehicles selected between 200 vehicles in the network. The maximum speed of vehicles is fixed to 30 m/s and the number of vehicular sources is fixed to 10 vehicles which are selected randomly to send CBR data at rate 5 packets/s to a node that is part of the wired network.

The performance of both schemes in term of packet delivery ratio, normalized routing overhead and average end-to-end delay increasing the number of vehicular gateway candidates in the network is shown in Fig. 2. From the figures, it can be seen that the network performance for both schemes increased while the percentage of gateway vehicles candidates is increased. The reason behind this is when the number of vehicular gateway candidates increases in the network, more the cost of finding and selecting one is low for vehicular sources, this is due to their availability in the network. The M-AODV+-VRGSA always outperforms the other scheme. M-AODV+-VRSA has the worst performance, this is because of the absence of a mechanism to select more appropriate gateway among neighboring vehicles which causes frequent handover and consumes more bandwidth in the network.

6.2   Varying Maximum Speed

We fixed the number of vehicles at 300, the number of vehicular sources selected randomly at 10 and the percentage of vehicle candidates to 50% (150 vehicular gateway candidates and 150 ordinary vehicles), to evaluate the performance of the proposed approach with increasing maximum speed.

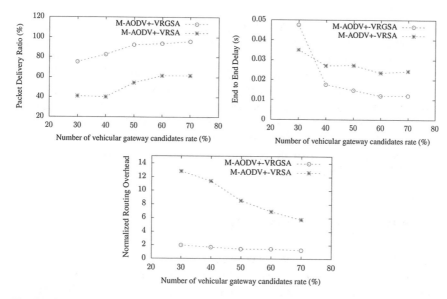

Fig. 2 Throughput, packet delay and overhead comparison varying number of dual interfaces vehicles in the network

Figure 3 illustrates the performance of both simulated mechanisms, in terms of packet delivery ratio, end to end delay, and normalized routing overhead respectively, for different maximum speeds of vehicles. As illustrated in packet delivery ratio and normalized routing overhead figures, the curves are almost stable in variation, this means that the impact of the increase of the maximum speed is not more important which shows the importance of using mobile infrastructures to connect VANETs

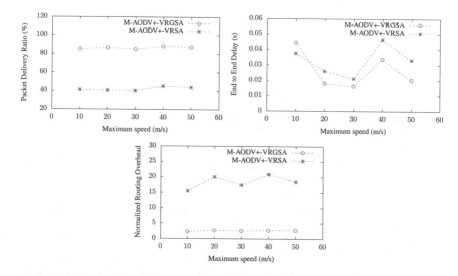

Fig. 3 Throughput, packet delay and overhead comparison under different maximum speed

nodes to the Internet. M-AODV+-VRGSA shows a 44.17% improvement in terms of packet delivery ratio and a 15.87% improvement in terms of normalized routing overhead over M-AODV+-VRSA. This outperformance is due to the fact that the mobility is taken into consideration to select more appropriate gateway which will result in the increasing of network connectivity that leads to the decrease of packets dropped in the networks. The variations in the average end to end delay against vehicle mobility speeds of M-AODV+-VRGSA and M-AODV+-VRSA are shown in end to end delay figure. As it is observed, both schemes show a significant impact on the end-to-end delay when the mobility speed increases, with slight improvement of M-AODV+-VRGSA.

7 Conclusion and Future Work

In this paper, we have designed a new proactive relay and mobile gateway selection scheme to select potential relays and gateways in the network, taking into account stability, mobility, path length, and quality of received signal. The objective is improving routing performance to provide Internet access in VANETs with high quality. Simulation results show that our scheme achieves better performance over a range of network performance measures.

In high vehicles density scenarios in VANET, one vehicle may be selected as a relay or mobile gateway for multiple vehicles, if the number of requests exceeds the service capacity of the vehicle, it might get overloaded, and eventually dropping the packets, increasing delay and enforcing retransmissions will be observed. In the future work, we will extend our approach by another overload control mechanism to avoid as much as possible, letting vehicles relays or mobile gateways becoming overloaded in the network.

References

1. Borcoci, E.: From vehicular Ad-hoc networks to internet of vehicles. In: NexComm 2017 Conference, April 23–27, Venice
2. Baras, S., Saeed, I., TabazaHadeel, A., Elhadef, M.: VANETs-Based Intelligent Transportation Systems: An Overview. Advances in Computer Science and Ubiquitous Computing, In book (2018)
3. Abada, D., Massaq, A., Boulouz, A.: Improving routing performances to provide internet connectivity in VANETs over IEEE 802.11p. Int. J. Adv. Comput. Sci. Appl. (IJACSA) **8**, 545–553 (2017)
4. Syfullah, M., Mun-Yee Lim, J.: Data broadcasting on Cloud-VANET for IEEE 802.11p and LTE hybrid VANET architectures. In: 3rd International Conference on Computational Intelligence and Communication Technology (CICT), 9–10 Feb. (2017)
5. Dharanyadevin, P., Venkatalakshmi, K.: Proficient Mmins algorithm in integratedvanet-4G milieu. Eur. J. Appl. Sci. IDOSI Publications **8**(4) (2016)
6. Sivaraj, R., Gopalakrishnay, A.K., Chandraz, M.G., Balamuralidharx, P.: QoS-enabled group communication in integrated VANET-LTE heterogeneous wireless networks. In: IEEE 7th

International Conference onWireless and Mobile Computing, Networking and Communications (WiMob) (2011)
7. Ucar, S. Ergen Sinem, S.C., Ozkasap, O.: Multi-hop cluster based IEEE 802.11p and LTE hybrid architecture for VANET safety message dissemination. IEEE Trans. Veh. Technol. **65**(4) (2015)
8. Garbiso, J., Diaconescu, A., Coupechoux, M., Leroy, B.: Auto-adaptive multi-hop clustering for hybrid cellular-vehicular networks. In: IEEE 20th International Conference on Intelligent Transportation Systems (ITSC), 16–19 Oct. (2017)
9. Rajendran, R.: The evaluation of geonetworking forwarding in vehicular Ad-Hoc networks. Master thesis in embedded and intelligent systems (2013)
10. Hussain, R., Rezaeifar, Z., Son, J., Bhuiyan, M.Z.A., Kim, S., Oh, H.: PB-MII: Replacing Static RSUs with Public Buses-Based Mobile Intermediary Infrastructure in Urban VANET-Based Clouds. Springer Science+Business Media, New York (2017)
11. Setiwan, G., Iskander, S., Kanhere, S.S., Jun Chen, Q.: Feasibility study of using mobile gateways for providing internet connectivity in public transportation vehicles. IWCMC (2006)
12. Namboodiri, V., Agarwal, M., Gao, L.: A study on the feasibility of mobile gateways for vehicular Ad-hoc networks. In: VANET '04 Proceedings of the 1st ACM International Workshop on Vehicular ad hoc Networks (2004)
13. Eltahir, A.A., Saeed, R.A., Mukherjee, A., Hasan, M.K.: Evaluation and analysis of an enhanced hybrid wireless mesh protocol for vehicular ad hoc network. EURASIP J. Wirel. Commun, Netw (2016)
14. Benslimane, A., Taleb, T., Sivaraj, R.: Dynamic clustering-based adaptive mobile gateway management in integrated VANET 3G heterogeneous wireless networks. IEEE J. Sel. Areas Commun. **29**, 559–570 (2011)
15. Benslimane, A., Barghi, S., Assi, C.: An efficient routing protocol for connecting vehicular networks to the internet. Pervasive Mob. Comput. J. Elsevier Publisher (2010)
16. Bhoyroo, M., Bassoo, V.: Performance evaluation of Nakagami model for Vehicular communication networks in developing countries. In: EmergiTech, IEEE, Balaclava, Mauritius (2016)
17. Killat, M., Hartenstein, H.: An empirical model for probability of packet reception in vehicular Ad Hoc networks. EURASIP J. Wirel. Commun. Netw. **1–12**, (2008)
18. Chen, S., Jones, H., Jayalath, D.: Effective link operation duration: A new routing metric for mobile ad hoc networks. In: International Conference on Signal Processing and Communication Systems, ICSPCS (2007)
19. Abada, D., Massaq, A., Boulouz, A.: Connecting VANETs to Internet over IEEE 80211p in a Nakagami fading channel. In: International Conference on Wireless Technologies, Embedded and Intelligent Systems (WITS). Publisher IEEE, Fez Morocco (2017)
20. Korner, U., Hamidian, A., Nilsson, A.: Performance of internet access solutions in mobile ad hoc networks. In: Wireless Systems and Mobility in Next Generation Internet, First International Workshop of the EURO-NGI Network of Excellence. Dagstuhl Castle, Germany (2004)
21. Abada, D., Massaq, A.: Improving relay selection scheme for connecting VANETs to Internet via IEEE 802.11p. Int. J. Comput. Appl. **132** (2015)
22. Ding, Z., Ren, P., Du, Q.: Mobility Based Routing Protocol with MAC Collision Improvement in Vehicular Ad Hoc Networks. Cornell University Library (2018)
23. Karnadi, F.K., Mo, Z.H., Lan, K.C.: Rapid generation of realistic simulation for vanet. In: IEEE WCNC (2007)
24. Simulation of Urban Mobility, http://sumo.sourceforge.net
25. The Network Simulator NS2, http://www.isi.edu/nsnam/ns/

Mobility Management: From Traditional to People-Centric Approach in the Smart City

Matteo Trombin, Roberta Pinna, Marta Musso, Elisabetta Magnaghi and Marco De Marco

Abstract The world's urban population is expected to double by 2050. As the planet becomes more urban, cities need to become smarter. Major urbanization requires new and innovative ways to manage the complexity of urban living; it demands new ways to target problems of over-crowding, energy consumption, resource management, and environmental protection. It is in this context that Smart Cities emerge not just as an innovative modus operandi for a future urban living but as a key strategy to tackle poverty and inequality, unemployment and energy management. At its core, the idea of Smart Cities is rooted in the creation and connection of human capital, social capital, and information and Communication technology (ICT) infrastructure in order to generate greater and more sustainable economic development and a better quality of life. Smart mobility represents one of the six dimensions of a smart city. It involves both environmental and economic aspects and needs both high technologies and virtuous people behaviors. Smart mobility is largely permeated by ICT, used in both backward and forward applications, to support the optimization of traffic flows, but also to collect citizens' opinions about livability in cities or quality of local public transport services. The aim of this paper is to analyze the Smart Mobility initiatives like part of a larger smart city initiative portfolio and to investigate about the role

M. Trombin (✉)
Università Telematica Internazionale UNINETTUNO, Rome, Italy
e-mail: matteo_trombin@libero.it

R. Pinna · M. Musso
Università di Cagliari, Cagliari, Italy
e-mail: pinnar@unica.it

M. Musso
e-mail: musso@unica.it

E. Magnaghi
Université Catholique de Lille, Lille, France
e-mail: elisabetta.magnaghi@univ-catholille.fr

M. De Marco
Università Cattolica, Milan, Italy
e-mail: marco.demarco@unicatt.it

© Springer Nature Switzerland AG 2020
M. Elhoseny and A. E. Hassanien (eds.), *Emerging Technologies for Connected Internet of Vehicles and Intelligent Transportation System Networks*, Studies in Systems, Decision and Control 242, https://doi.org/10.1007/978-3-030-22773-9_11

of ICT in supporting smart mobility actions, influencing their impact on citizen's quality of life and on the public value created for the city as a whole.

Keywords Mobility management · Smart city · People-centric approach · Mobility as a service (MaaS) · Technology strategic dimensions

1   Smart City Definitions: An Overview

In the last decade in literature, there is increasing interest in the concept of Smart City. The Smart City is a new "Intelligent" city paradigm, which combines the widespread use of new communication technologies, mobility, the environment, and energy efficiency, in order to improve the quality of life and meet the needs of citizens, businesses and institutions. The current reference scenario is characterized by some trends that influence and will influence the development of the cities of the future, such as a sustained growth of the population, combined with a continuous increase in the level of urbanization (78% of the population of advanced developing countries live in centers urban); a significant environmental impact of cities that consume 75% of energy and are responsible for 80% of CO_2 emissions worldwide; a globalization that exposes the urban realities to a continuous growth with consequent problems of overcrowding, congestion, inadequacy of the transport systems. For these reasons, in recent decades, the city has acquired greater centrality in the economic, environmental and social development process and has become a focal point of the economic policies and strategies of international bodies and legislators. The Smart City concept is always with greater emphasis indicated as a strategic solution to the problems associated with the irreversible urban agglomeration process Despite the vast literature on smart cities there are more than thirty-six definitions of the term. This is because the idea of Smart City is relatively new and evolving, and the concept is very broad. However, the literature does not provide a unified a definition of the construct that is (a) inclusive of the present definitions, and (b) extensible to accommodate the evolution of the construct (See Table 1). Many definitions of the Smart City focus almost exclusively on the fundamental role of ICT in linking city-wide services. For example, one suggestion is that a city is smart when: 'the use of ICT [makes] the critical infrastructure components and services of a city—which include city administration, education, healthcare, public safety, real estate, transportation, and utilities—more intelligent, interconnected, and efficient'. Similarly, another approach states, 'We take the particular perspective that cities are systems of systems and that there are emerging opportunities to introduce digital nervous systems, intelligent responsiveness, and optimization at every level of system integration.'

The first attempts to define the concept were focused on the smartness provided by information technology for managing various city functions [15, 24, 36]. By this point of view, some researchers have stressed technology and infrastructure as the main components. The use of information technology has been considered as a key

Table 1 Main definitions of smart city

Technology-centered definitions	The use of ICT [makes] the main components of a city's infrastructure and services more intelligent and efficient: public administration, school, health system, security, transport, real estate, distribution networks of water, gas, electricity	Washburn and Sindhu (2009)
	Cities [must be seen as] systems of systems and there are increasing opportunities to insert digital nervous systems, intelligent responsiveness, and optimization at every level of system integration	MIT (2013)
	In an Intelligent City the networks are interconnected, support and feed each other positively and in such a way that: the collection of data can allow analysis and distribution of information about the city at all times to optimize effectiveness and efficiency in pursuing competitiveness and sustainability; it is possible to communicate and share information in every part of the city thanks to common usage definitions and standards, such as to make it easy to re-use; we can act multi-functionally, so as to provide solutions to multiple problems in a holistic perspective for the city	Copenhagen Cleantech Cluster (2012)
Overall definitions	A city is intelligent when investments in social and human capital and in infrastructure for traditional and modern communication feed sustainable economic growth and a high quality of life, with careful management of natural resources, through participatory government processes	Caragliu, Del Bo and Nijkamp (2009)
	Talking about Smart City means] leveraging interoperability within and between different political areas of a city (transport, public security, energy, training, health, and development). Smart City strategies require innovative ways of interacting with stakeholders, resource management and service provision	Nam and Pardo (2011)
	A city can be defined as 'intelligent' when investments in social and human capital and in infrastructure for traditional and modern communication feed sustainable economic growth and a high quality of life, with careful management of natural resources, through processes of participated government	[33]
	Any model to be appropriate for the Smart City must also focus on the ability to be "smart" of its citizens and communities, on their well-being and on their quality of life and must encourage the processes that make cities important for people and they could support very different and sometimes conflicting activities	Haque (2012)
Working definition of the EU research	A Smart City is a city that seeks to address issues of public interest through ICT solutions based on municipal multi-stake-holder partnerships	Mapping Smart Cities in EU (2014)

factor in the smartness of a city since it can sense, monitor, control and communicate most of the city services like transport, electricity, environment control, crime control, social, emergencies, etc. [7, 23, 8, 1]. While information technology can make a city smart (or smarter), the city itself is an entity with multiple stakeholders seeking diverse outcomes. Other studies [2, 16], suggest to include the outcome of the Smart City such as sustainability, quality of life, and services to the citizens. For example, [33] in their definition tend to balance different economic and social factors with an urban development dynamic. For the authors "a city may be called Smart when investments in human and social capital and traditional and modern communication infrastructure fuel sustainable economic growth and a high quality of life, with a wise management of natural resources, through participatory governance". Murgante and Borruso (2015) warned that cities, in the rush of being considered part of the "Smart umbrella", can be susceptible to ignore the importance of becoming sustainable and if they focus solely on improving technological systems, they can easily become obsolete. The proposed unified definition integrates the two aspects. Chourabi et al. [6] propose a framework that attempts to incorporate sustainability and livability issues, as well as internal and external factors affecting smart cities. They identify eight factors that, based on the literature at that time, were considered fundamental to the comprehension of smart city initiatives and projects. They include management and organization, technology, governance, policy, people and communities, the economy, built infrastructure, and the natural environment. The same spirit of providing a more integrated perspective of smart cities prevails in [25] who present a taxonomy of domains, hard and soft and they grouped the key elements into six categories: natural resources and energy; transport and mobility; buildings; living; government; economy and people. Giffinger et al. [12] also conceive a framework, based on the literature, for ranking smart medium-sized cities in Europe. They conceive a smart city as one that would excel, in a forward-looking way, in six characteristics: smart economy, smart people, smart governance, smart mobility, smart environment, and smart living. Similarly, [18] proposed a framework based on the concept of the Triple Helix that relates to university, industry, and government. They identify five clusters of elements in their analysis: smart governance, smart human capital, smart environment, smart living, and smart economy. The indicators for the dimensions in the framework were designed using a focus group and experts in different disciplines to allows a future classification of smart city performance and the relations between components, actors, and strategies. In this study, we used the following definition of Smart City as "a city seeking to address public issues via ICT-based solutions on the basis of a multi-stakeholder, municipally-based partnership". This definition points out that a Smart city is a construct composed of two dimensions: City and Smart. The term City is defined by its stakeholders and the outcomes, like as Sustainability, Quality of Life, Equity, Livability, and Resilience. One of the pivotal drivers of this new paradigm is qualitative because it encompasses quality of life, environmental sustainability and the higher needs and exigencies of the modern human being. The Stake-holders in a city include its Citizens, Professionals, Communities, Institutions, Businesses, and Governments. The effects on 'citizens' quality of life', 'communities' equity', 'businesses' resilience', and other possible combinations of Stakeholder and Outcome, defines the smartness of a city. The term smart identifies the idea of City that ensures a better quality of life through the use of ICT. Broadly speaking, smart cities are cities well performing in the following six aspects: smart economy, smart people, smart mobility, smart environment, smart living and smart governance [12, 19].

2 Main Characteristics of a Smart City

The definition of Smart City points out that gives plenty of initiatives in the technical and policy environment within socio-economic dynamics potentially give rise to a wide variety of Smart City characteristics. European Smart City Project proposed six characteristics (see Fig. 1). A Smart City, to be classified as such, is a city that has at least one initiative that addresses one or more of the following six features: smart governance, smart economy, smart mobility, smart people, smart living, smart environment.

The concept of Smart Governance refers to the idea of an administration, that should be able to manage not only the development and transformation plan in Smart City but also the interactions that connect public, private and civil society organizations, to ensure efficient operation and effective in the city. Smart objectives include the transparent operation of administration with open data systems, with

Fig. 1 Main characteristics of a smart city

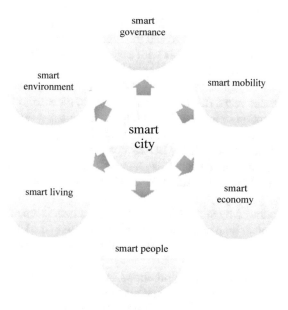

the use of ICT, the participation of citizens in the decision-making process and the "co-creation" of E-services.

Smart Economy means the adoption of e-business and e-commerce services through the use of ICT-enabled systems that favor both internationalization processes but also knowledge sharing and creativity enhancement. Investment in the knowledge economy is a precondition for governing the process of transformation towards Smart Cities, which need to promote a synergic system between a private company, public bodies, research institutes capable of acting as a stimulus both for companies and for businesses for every single citizen.

The concept of smart environment recalls the characteristics of an intelligent environment such as intelligent energies including renewable energies, the enabling of ICT for energy networks, measurement systems, pollution control and monitoring systems, the restructuring and rationalization of the building, protection and management of urban green areas and the reclamation of disused areas, as well as energy efficiency and environmental sustainability, differentiated collection and reuse initiatives, public street lighting, the systems of drainage and water systems.

The transition to a smart city requires that public governance pay the utmost attention to creating the conditions that allow the formation of smart people, i.e. people with IT skills, ICT technology experts, people who have access to education and training. The central idea is that within an inclusive society the creativity of citizens, experts in shared efforts must be enhanced, encouraging innovation and interactivity.

The concept of Smart Living recalls the idea of living in health and safety in a culturally vibrant city, with different cultural structures, and includes homes and accommodation in good quality accommodation facilities. A Smart city that wants to engage in this dimension must undertake actions to promote its tourism image through innovative tools such as intelligent presence on the web and virtualization of its cultural heritage and traditions. This means networking a "common good" for its citizens and visitors with easy-to-use theme itineraries and maps of the city.

Finally, the set of transport and logistics systems supported and integrated by the ICT constitute the dimension of Smart Mobility. The constant growth of mobility in recent years, due to population growth, globalization and urbanization, require the adoption of intelligent mobility management systems.

From what emerges from the study "Mapping Smart Cities in EU" in 2011 [10], of the 468 cities in the EU-28 area with more than 100,000 inhabitants mapped in the aforementioned study, as many as 240 (51% of the total) had an initiative linked to one of the six features of Smart City, thus being able to be classified as such. As can be seen in Fig. 2, the feature that encompasses most initiatives is the Smart Environment, followed by Smart Mobility. Moreover, Italy is among the countries with the highest number both in absolute terms (together with the United Kingdom and Spain) and in percentage terms (with Austria, Denmark, Norway, Sweden, Estonia, and Slovenia) of Smart Cities.

Moreover, the report shows that big cities tend to have a higher maturity level of the projects since they already have kicked off or put into practice initiatives of a smart city. With reference to the overall number of initiatives involving EU

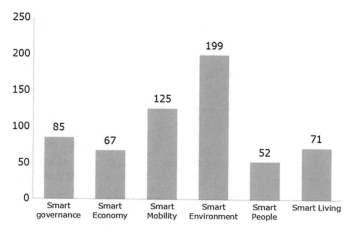

Fig. 2 The number of smart cities in the EU presenting the six characteristics. *Source* European Parliament, Mapping Smart Cities in the EU, 2014

countries, in only 27% of cases the Smart Cities strategy or policy was defined, in 25% there is the availability of a project plan not yet started, in 20% there was a pilot experimentation of Smart cities initiatives and only in 28% was a "Smart" initiative completely completed.

3 Smart Mobility

Smart Mobility is one of the topics regarding Smart City implementation. It is, however, a crucial topic, impacting on several dimensions of the smart city, on different aspects composing the citizens' quality of life and regarding all the potential stakeholders expecting benefits from the smart city implementation. The concept of a smart city implies that of smart mobility, which is a set of transport and logistics systems supported and integrated by the ICT [13, 4]. The term embodies a series of elements: technology, mobility infrastructures (parking lots, recharging networks, signage, vehicles), mobility solutions (including new mobility models) and people (Arena et al. 2013).

The UN Climate Change Targets, adopted in Paris during the World Climate Change Conference 2015, are still valid and are far from being reached. The particulate matter problem often exceeds the agreed limit values many times over. The volume of traffic in larger and medium-sized metropolitan regions regularly exceeds any reasonable and sustainable level. In addition, there are insufficiently coordinated traffic connections and an inefficient ticket and booking management system. Of course, there is also the question of the so-called first and last mile, i.e. how to get from A to B with minimum impact and highest efficiency. In this regard, concrete solutions are needed. Based on current scenario projections, a radical transformation of transport

Table 2 Background principles of smart mobility

Flexibility	Multiple modes of transport allow those who move to choose which of these is the best in a given context
Efficiency	The traveler is able to arrive at their destination with minimum effort and in the shortest possible time
Integration	The complete journey is planned without taking into account which means of transport are used
Clean technologies	From the vehicles that cause pollution, we move towards those with zero emissions
Safety	The rate of dead and blessed people dramatically shrinks
Accessibility	Everyone should access different forms of smart mobility

Source Our elaboration

systems is required and will become a key policy challenge. Transport transformation and innovation scenarios currently focus mainly on fuel efficiency, fuel substitution, and end-of-pipe carbon capture as levers for decarbonization. Future efforts need to focus on the combined and synergetic effects of integrating urban energy, infrastructure and mobility systems including via modal-shift measures, expansion of public transport options, and sustainable land use governance.

The ultimate goal of introducing smart mobility in our cities is to reduce traffic, reduce pollution, create intelligent and uninterrupted flows, and strengthen economies of scale to promote mobility accessible to all. Smart mobility does not just mean alternative forms of transport. It is a broader and more complex phenomenon in that it aims to offer a mobility experience that is flexible, integrated, safe, accessible and convenient so as to contribute to improving the quality of life (Table 2). At the core of Smart Mobility there is the concept of integration: between the various transport systems, between infrastructures and users. To achieve integration, an integrated mobility platform is needed, which is attractive to both users and providers of mobility.

This platform enables the integration of different means of transport, ideally combining information, booking, ticket creation and billing with the mobility providers' respective loyalty programs to create an intelligent mobility application through which traffic flows can be routed.

3.1 Main Innovations for Smart Mobility

Smart mobility can be seen as a combination of four domains. Firstly, smart mobility is about vehicle technology: power trains, electric car technology, fuel technology, autonomous automation, driver assistance systems, but also new types of bicycles. Secondly, smart mobility is about Intelligent Transport Systems: cooperative adaptive cruise control, traffic management, connected automated driving, platooning of trucks. Thirdly, smart mobility is about data: travel information, logistics planning, advanced IT systems for matching supply and demand, big data solutions. And finally, smart mobility is about new mobility services: seat management, car sharing, ride sharing, connecting transport modes, new cycling systems. These four domains—vehicle technology, ITS, data, new mobility services—broadly define the current scope of smart mobility that finds its origins in a combination of innovation in transportation, infrastructures, and services. The strategic technology dimensions related to smart mobility in the next future will be related to Autonomous things, Augmented analytics, AI-driven Development, Digital Twins, Edge Computing, Immersive experience, Blockchain [22].

3.1.1 Autonomous Things

The Internet of Things and the spread of 5G will have a significant impact on the world economy and will also profoundly transform the mobility sector. Which will be characterized by the optimization of the movements of people and goods and, in particular, by the introduction of technologies that will lead to automation in the road, rail and air traffic [17]. In facts, the impact of the internet on mobility is already evident and has led to considerable advantages in terms of traffic (for example smartphones and GPS that indicate routes and roads with less traffic in real time), travel (access to information on timetables and locations public transport) and shared mobility (optimization of vehicle and vehicle movements). Nevertheless, we are approaching a further change of paradigm, which brings with it some characteristics that will have a profound impact on people and society, with services radically different from those we know today. The meeting of the characteristics of 5G (in particular large capacity and very low latency) with the spread of smart sensors in the environment, in objects and in wearable devices, is transforming the Internet into a network capable of progressively connecting all the machines and robots allowing them to perform actions. In the mobility sector, the network is being transformed from an information support technology for drivers and passengers to a real entity capable of autonomously controlling the means of transport. The machines, although they are not really capable of making autonomous decisions, will act on two inputs: behind direct impulse of human beings through multiple interfaces, among which the vocal one grows - and on the basis of complicated algorithms able to process in real time an immense amount of information and to guide the machines in relation to external inputs and patterns of pre-established behavior [29]. Through the cloud,

AI and IoT technologies, cities can harness the power of real-time intelligence for monitoring, anticipating and managing urban events, from traffic congestion and flooding to utility optimization and construction [27].

3.1.2 Augmented Analytics

National road traffic data and data flows from public transport cards, smartphones, navigation systems, connected vehicles, and social media, provide a growing collection of information that provides insight into the behavior of road users, experts in the field of information technology and mobility in the magazine for network management say in traffic and transport. As a result, it is possible to better understand what drives people to move, what their motives are for choosing a car or public transport, which routes they prefer when they are in a hurry, and to what extent they adjust their driving behavior accordingly. Thanks to machine learning and AI it is possible to automate data analysis processes making data analysis easy even for non-experts, thus accessibility becomes a relevant factor for citizen's engagement. Through increased analysis it is possible to identify the hidden models, removing subjective and even unconscious bias.

3.1.3 AI-Driven Development

Artificial intelligence enables a range of applications that make mobility more comfortable, accessible, efficient and resource-efficient in short, smarter. The following overview shows examples, introduced by the members of the focus group Intelligent Mobility. They illustrate that the mobility sector is an important area of application for AI solutions and that the German mobility industry has already recognized the potential of artificial intelligence and is beginning to deploy it. Many actors are testing AI-based technologies in tests and research projects that are supporting or enabling Intelligent Mobility (see e.g. Voda et al. [38]). Autonomous objects leverage AI to interact more spontaneously with their environment and to perform tasks that are traditionally performed by human beings. As a consequence, a general framework emerges, by which every application, service, and object could potentially embody a form of AI to either fully automatize or improve their own process.

3.1.4 Digital Twins

The model of digital twins applied to IoT for smart mobility makes use of Big Data Analytics and AI. The model is linked with real-time physical situations to be monitored and controlled. Some cities, like Rotterdam (which focuses upon the numerous bridges to open for water and road traffic), have been creating a "digital twin" for the city, which will act as a platform for a new era of digital city applications. Sensors and data-streams around the city feed into the model meaning many digital elements are updated in real-time, such as the movement of people or vehicles, which can be incorporated into the system. In this scenario, a digital twin would contain detailed design information about the road layout and construction, drainage systems, gantries, and signage. It would combine this with predicted traffic flow models, incident management plans, and, once operational, live and projected traffic flow data [21]. The intelligence unleashed by the digital twin enables cities to improve their physical and social urban environment. It also offers a more democratic solution, which is evidenced by the increasingly wide variety of locally-designed applications that greatly improve the services to communities.

3.1.5 Edge Computing

With edge computing, resources are distributed at the edge of the mobile network, i.e., in base stations, allowing a dynamic resource allocation. This approach significantly reduces delay for computation of tasks offloaded from users' devices to cloud and reduces the load of backhaul. Comparing to state of the art approaches, the proposal leads to a reduction of the task offloading delay between 10 and 66% while energy consumed by user's equipment is kept at a similar level. The proposed algorithm also enables higher arrival rate of the offloading requirements [28]. Beyond the obvious operational cost savings, low latency and resilience benefits, the presence of edge application platforms also opens up the possibility of developing a limitless range of applications, particularly street lighting, sensor monitoring, connected parking meters, etc. In particular, traffic management is an ideal application for edge computing technologies. Also, smart transportation dramatically improves with the aid of edge computing; it reduces traffic accidents with connected infrastructure, data analytics, and machine learning that can optimize traffic systems and identify high-accident intersections. Machine learning, when deployed on the edge devices, will enable connected traffic systems to engage with and react to both manned and autonomous vehicles. Machine learning tools collect traffic data from the Internet of Things (IoT) sensors that are embedded in the roads and in the traffic lights, historical surveys, radar images, etc. But edge computing also provides a computing platform that not only benefits cities as a whole but also citizens directly [30]. To sum this up, the benefits of edge intelligence with machine learning reduce the data deluge and minimize the delay in communication, which ultimately reduces the cost [31].

3.1.6 Immersive Experience

According to Gartner (2018), by 2028, conversational platforms (ranging from virtual personal assistants to chatbots) and technologies such as augmented reality (AR), mixed reality (MR) and virtual reality (VR) will lead to new immersive experiences able to increase productivity. Overhaul and servicing by means of unmanned vehicles (UAV), the use of augmented reality for operator training and remote assistance represent the kind of services that will be 'augmented' and profoundly modified by the use of immersive experience applied to mobility.

3.1.7 Blockchain

Blockchain technology is invading every aspect of life and has been finding an application to almost every field. Many efforts are devoted to applying the blockchain technology to mobility in order to improve decisive aspects and services [35, 20, 34, 26]. Though, one of the most interesting contributions has been lately worked out by Scekic et al. [32], which envisages the idea of a blockchain-supported smart city platform for social value co-creation and exchange. Value co-creation is actually crucial to a successful implementation of smart mobility and performing citizen's behavior towards the acceptance and adoption of new technologies.

3.2 The Model "Mobility as a Service" as Engagement Tool

Mobility as a service (MaaS) is an interesting transition in mobility. MaaS makes it possible to get in at the place of departure and get out at the place of arrival via a smoothly organized service (mobility as a service) [14]. It is less relevant for the traveler who organizes the mobility. He uses an app to communicate his destination, after which the system proposes an ideal travel time and means of transport based on parameters such as arrival, price and predicted traffic jams. The app can then automatically arrange the entire journey, including buying train tickets and reserving shared cars, for example. This model, which implies a change of mind by going beyond the concept of property, brings significant advantages to users, managers, operators, and organizations, who will all benefit from shrinking costs, increased driving safety and reduction of environmental impact. Users will benefit from optimized transport, cheaper and without technological barriers, with real-time information adapted to their preferences and needs. All this will be possible thanks to the technologies that are enabling the so-called shared economy; too big data and solutions for urban and long-distance inter-mode. For example, from a single application, the user can configure his or her profile, choose different means of transport and pay at the end of the month based on usage, as well as benefit from "rewards" for more sustainable choices. On the operators' front, the new reality will allow for a more intelligent, intermodal and optimized management of their services. The operators of buses, subways and trains will have at their disposal an integrated management system of the routes, centralized and connected with the information of travelers and traffic, in which automatic learning and big data will make it easier to resize routes in real time and create personalized routes for users of other means of transport. Traffic managers and motorway concessionaires will evolve towards a mobility-as-a-service model for car users, which will include solutions for access to certain routes, modular prices, and payments based on road conditions, per day, time and occupation. Moreover, the connected and/or autonomous car developed thanks to the IoT and to the new secure mobile communications, will radically change the user experience, which will have access to all the information and assistance services available for safer driving. The optimization of costs in infrastructure maintenance, the traceability of goods and the optimization of the so-called "last mile" and the final delivery, represent the main challenges indicated respectively by the owners of the infrastructures and by the logistics operators. The latter will have integrated freight management platforms with total traceability, based on blockchains and smart contracts, connected to traffic data in real time. This will simplify the distribution, which will have a lower environmental impact by optimizing the load on trains and trucks. The maintenance of transport infrastructures will be increasingly predictive, starting from the infrastructure design itself in BIM (Building Information Model), up to the collection of relevant data through big data to establish indicators and models that, supported from artificial intelligence, they will facilitate decision-making.

The scenario with "mobility as a service" will only see the car as a vehicle within the fleet of mobile service providers to meet subscriber's requests through an integrated and fully co-modal transport supply. Urban and interurban journeys will mostly be made on non-vehicles ownership: cars will grow in sharing and individual ownership will be less important. A driver will use several vehicles in a short time and each vehicle will be used by a multiplicity of drivers. Specialization (urban, extra-urban vehicles, suitable for fog, with different fueling systems, etc.) and intensity of use, with the reduction of long stops of owned vehicles, will justify the cost of designed vehicles different and with sophisticated onboard systems.

4 Sustainable Urban Mobility Plans: The People-Centric Approach

Mobility management is a complex activity because planners are called to take on many, often conflicting, requests expressed both at local and European level, which increasingly require the search for alignment with European policies for the fight against climate change and energy efficiency objectives. For this reason, since 2009 the EU has promoted the adoption of Sustainable Urban Mobility Plans (SUMPs) as a planning tool of new conception, able to face the challenges posed by transport and the critical issues affecting urban areas in a more integrated and sustainable perspective. The European Commission [9] defines a SUMP as a plan aiming to improve the accessibility of urban areas and providing high-quality and sustainable mobility and transport to, through and within the urban area. It focuses on the needs of the "functioning city" and its hinterland rather than on a municipal administrative region. The new concept introduced by the SUMPs particularly emphasizes the involvement of citizens and stakeholders, on the coordination of policies and planning tools between sectors (transport, urban planning, environment, economic activities, social services, health, safety, energy, etc.), between institutions, between different

Table 3 Main differences between the traditional and the new approach

Traditional approach	People-centric approach
Traffic is at the core	The individual is at the core
Goals: the capacity of traffic flow and speed	Goals: accessibility and quality of life, sustainability, financial affordability, social equality, health
Modal focus	Development of different transport modality
Infrastructural focus	Integrated solutions to generate effective and cost-efficient solutions
Expert-driven planning	Interdisciplinary workgroups

levels within and across the territory and between neighboring institutions. It aims at building on existing planning practices and ensuring integration, as well as participation and evaluation principles. People are placed at the core of SUMPs; whether it concerns commuters, business people, consumers, customers or any other role, provided a SUMP means "Planning for People" [11]. This people-centric approach is one of the main differences from more traditional transport planning, which tend to focus on traffic and infrastructure, rather than on people and their mobility needs (see the Table 3).

In 2013, this resulted in the publication of guidelines for the development and implementation of Sustainable Urban Mobility Plans (SUMPs). These guidelines provide local authorities with a structured approach on how to develop and implement strategies for urban mobility based on thorough analysis of the current situation, combined with a clear vision for sustainable development of the urban and neighboring areas under consideration. Thereby, SUMPs can help cities make efficient use of existing transport infrastructure and services and ensure a cost-effective deployment of the proposed measures. A SUMP focuses on the fulfillment of individual mobility needs, by leveraging a participatory approach that involves the active involvement of citizens and other stakeholders from the beginning of (and during) the development and implementation of the entire process.

5 Mobility Management: The State-of-the-Art in Italy

The European Platform of Mobility Management (EPOMM) defines Mobility management as "a concept for promoting sustainable transport and dealing with the question of car use by modifying the habits and behavior of travelers. The core of this mobility management is formed by "soft" policy measures such as information and communication, the organization of services and the coordination of activities of the various partners". More and more European cities are turning to this type of measure to provide residents with information on all sustainable modes of transport (e.g. public transport, bike, car sharing, etc.) and to eliminate the physical and psychological barriers that limit their mobility choices. It has been noted that frequently the reasons for the non-use of public transport consist of the scarce knowledge of the real options and the bad reputation of this mode of transport. By informing people about the alternatives to the use of private cars, their mobility behavior can be influenced and a modal shift towards more sustainable modes of transport can be stimulated, thus helping to reduce car traffic and its negative impacts.

Italy is a country with one of the most important European and world-wide historical heritage. However, this cultural heritage is forced to coexist with one of the highest motorization rates among all European countries: the number of cars per 100 inhabitants keeps increasing and it reached 61.6, a very high figure compared with the European average (49.1). Rome, capital of Italy as well as the main historical city of the country, confirms the national average, reaching about 61.3 cars per 100 inhabitants. Focusing attention on passenger transport, it now seems certain that the relative problems do not lie only in the lack of infrastructure or services but also in a widespread cultural attitude that makes the use of the car more preferable to the citizen than any other mode of transport. Such scenario urged the Italian Ministry of Infrastructures and Transport to adopt in 2000 a specific Decree containing the national guidelines on Sustainable Urban Mobility Plans, with the aim to improve the supply system (also through the implementation of computing and digital technologies). Most of all though, its main aim was at controlling the mobility demand. Indeed, a decree of the Ministry of Environment introduced the profile of mobility manager for companies and institutions with over 300 employees in a local unit or a total of over 800 employees in several local units. The aim of the decree is to engage business and workers in identifying alternatives to the use of the private vehicle. The mobility manager is responsible for optimizing systematic trips between home to work of the staff, introducing new forms of environmentally sustainable mobility to address current issues of air pollution and traffic congestion. More precisely, the role of the mobility manager at the city level is primarily to organize training and courses for company mobility managers and to provide them with technical support for the preparation of home-to-work mobility plans, by providing software tools and/or informational materials for the preparation of these plans. Mobility managers at the city level, however, play also an important role for the local community, promoting sustainable mobility and sensitizing the community through the organization of events, conferences and citizens' involvement in projects dedicated to mobility. The reference contact person for the company Mobility Manager is the mobility managers of the area who has an important coordination function and acts as an intermediary between all the different parties involved. These collect the needs of the individual Mobility Managers and develop strategies aimed at managing home-work mobility as a whole by promoting the measures implemented and using strategies to involve and participate, citizens, workers, and employers use to identify and manage the alternative options. Its main objective is to create the framework conditions for reducing the propensity to use the car.

From the analysis of the data provided by Euromobility (the Association of Mobility Managers) and material collected and the observation of various cases of Mobility Management initiatives identified on the Internet it emerges that, despite being a young reality that of Mobility Management in Italy has developed a strong dynamism around the theme of sustainable mobility. The picture that emerged is very comforting as the websites identified were dedicated to this issue and above all, there was an increasing need to create ad hoc websites for Mobility Management. In fact, municipal administrations are increasingly inclined to include in their websites dedicated sections or to prepare thematic portals on mobility management in which, even if no concrete activities have been implemented, the figure of the mobility manager is presented and the various projects carried out are presented or that will be carried out. Moreover, many companies have used this tool as a showcase through which to make them known to users and to public opinion. An example is that of the ASL (Local Health Company) n. 4 of Senigallia, which has published its own PSCL by dedicating a section to corporate mobility management. Furthermore, several initiatives have been identified to produce a sustainable and safe transport offer related to home-school travel. This data can also be read as an educational attempt aimed at younger people to stimulate awareness of the impact that transport systems have on the environment and on quality of life.

Furthermore, it was possible to observe that Italian Mobility Management was substantiated above all in the preparation of home-work travel plans. The measures adopted mainly concerned the promotion of local public transport, the development of cycling mobility, the dissemination of carpooling and the development of corporate transport services. Only the province of Milan proposed alternative measures to the traditional ones, such as the promotion of LPG and experimentation with electric vehicles. However, in recent years more attention is also being paid to other "poles" of attraction, such as schools, fair areas, commercial areas, and hospital centers. Among the initiatives involving the journey from home to school, the area of the Municipality of Milan was of particular interest, as it is preparing an agreement between the Municipality and the Region to establish the figure of the Regional Mobility Manager for the school. In the Municipality of Bologna, on the other hand, the creation, within the project for a child-friendly city, of educational workshops to raise awareness and raise awareness of the moving house-to-work plans that are being set up is active.

6 Conclusion

The continuous growth of urban transport and its environmental effects urgently requires the need to strengthen the actions and instruments capable of influencing the demand for mobility. The success of these policies would, in fact, allow the achievement of important objectives set by the European directive on air quality, in particular for particulates that exceed the limits in many urban areas, and to contain the growth of carbon dioxide emissions thus facilitating achieving the objectives of the Kyoto Protocol. However, an approach to mobility often emerges that is still too often characterized by more emotional than rational factors and by a generic culture. Conversely, mobility management requires a scientific approach, a global vision of the phenomenon that allows evaluating every single initiative that is intended to be taken in the logic of economic evaluation of costs and benefits, and in that of the ability to evaluate and communicate impact on the community. The mobility manager and the mobility manager area could represent a key figure in the analysis of the flows of citizens' mobility, as they link the world of work with that of decision makers and operators of collective mobility.

References

1. Akhras, G.: Smart materials and smart systems for the future. Can. Mil. J. **1**(3), 25–31 (2000)
2. Anthopoulos, L. G.: Understanding the smart city domain: A literature review. In *Transforming city governments for successful smart cities*, pp. 9–21. Springer, Cham (2015)
3. Arena, M., Cheli, F., Zaninelli, D., Capasso, A., Lamedica, R., and Piccolo, A.: Smart mobility for sustainability. In AEIT Annual Conference 2013 (pp. 1–6). IEEE. (2013)
4. Benevolo, C., Dameri, R. P., and D'Auria, B.: Smart mobility in smart city. In Empowering Organizations, pp. 13–28. Springer, Cham (2016)
5. Caragliu, A., Del Bo, C., Nijkamp, P.: Smart cities in Europe. J Urban Technol **18**(2), 65–82 (2011)
6. Chourabi, H., Nam, T., Walker, S., Gil-Garcia, J. R., Mellouli, S., Nahon, K., Pardo, T.A., Scholl, H.J.: 2012 45th Hawaii International Conference on Understanding Smart Cities: An Integrative Framework, System Science (HICSS), pp. 2289–2297. IEEE (2012)
7. Cocchia, A.: Smart and digital city: A systematic literature review. In: Dameri, R.P., Rosenthal-Sabroux C., (eds.) Smart City: How to Create Public and Economic Value with High Technology in Urban Space, p 13–43. Springer International Publishing, Cham (2014)
8. Debnath, A. K., Chin, H. C., Haque, M. M., and Yuen, B.: A methodological framework for benchmarking smart transport cities. Cities. **37**, 47–56 (2014)
9. European Commission.: Sustainable Urban Mobility: European Policy, Practice and Solutions (2017)
10. European Parliament.: Mapping Smart Cities in the EU (2014)
11. European Commission.: Guidelines. Developing and implementing a Sustainable Urban Mobility Plan (2014)
12. Giffinger, R., Fertner, C., Kramar, H., Kalasek, R., Pichler-Milanović, N., Meijers, E.: Smart Cities - Ranking of European medium-sized cities. Cent. Reg. Sci. (2007)
13. Jeekel, J.F.: Smart Mobility and Societal Challenges: An Implementation Perspective. Technische Universiteit Eindhoven, Eindhoven (2016)
14. Karinsalo, A., Halunen, K.: Smart contracts for a mobility-as-a-service ecosystem. In: 2018 IEEE International Conference on Software Quality, Reliability and Security Companion (QRS-C), Lisbon, pp. 135–138 (2018). https://doi.org/10.1109/qrs-c.2018.00036
15. Komninos, N., Pallot, M., Schaffers, H.: Special issue on smart cities and the future internet in Europe. J. Knowl. Econ. **4**(2), 119–134 (2013)
16. Lee, J., Kao, H. A., & Yang, S.: Service innovation and smart analytics for industry 4.0 and big data environment. Proc. Cirp. **16**, 3–8 (2014)
17. Li, S., Da Xu, L., Zhao, S.: 5G internet of things: A survey. J. Ind. Inf. Integr., (10), Elsevier (2018)
18. Lombardi, P., Giordano, S., Farouh, H., and Yousef, W.: Modelling the smart city performance. Innov.: Eur. J. Soc. Sci. Res. **25**(2), 137–149 (2012)
19. Lombardi, P., Giordano, S., Farouh, H.,and Wael, Y.: An analytic network model for Smart cities. In Proceedings of the 11th International Symposium on the AHP, June, pp. 15–18. (2011)
20. López, D., Farooq, B.: A blockchain framework for smart mobility. In 2018 IEEE International Smart Cities Conference (ISC2). IEEE. pp. 1–7. (2018)

21. Malone, A.: How a digital twin could transform road network delivery (2018). https://www.wsp.com/en-GB/insights/how-a-digital-twin-could-transform-road-network-delivery

22. Mathew, E., Al-Mansoori, S.: Vision 2050 of the UAE in Intelligent Mobility, In 2018 Fifth HCT Information Technology Trends (ITT). IEEE. pp. 213–218. (2018)

23. Murgante, B., Borruso, G.: Smart cities in a smart world. In *Future City Architecture for Optimal Living*, pp. 13–35. Springer, Cham, (2015)

24. Nam, T., Pardo, T.A.: Conceptualizing smart city with dimensions of technology, people, and institutions. In Proceedings of the 12th annual international digital government research conference: digital government innovation in challenging times. pp. 282–291. ACM. (2011)

25. Neirotti, P., De Marco, A., Cagliano, A. C., Mangano, G., and Scorrano, F.: Current trends in Smart City initiatives: some stylised facts. Cities. **38**, 25–36 (2014)

26. Orecchini F., Santiangeli A., Zuccari F., Pieroni A., Suppa, T.: Blockchain technology in smart city: A new opportunity for smart environment and smart mobility. In: Vasant, P., Zelinka, I., Weber, G.W., (eds.) Intelligent Computing and Optimization. ICO 2018. Advances in Intelligent Systems and Computing, vol. 866. Springer, Cham (2019)

27. Pendleton, C.: IoT for smart cities: New partnerships for Azure Maps and Azure Digital Twins (2018). https://azure.microsoft.com/sv-se/blog/iot-for-smart-cities-new-partnerships-for-azure-maps-and-azure-digital-twins/ (retrieved 13 March 2019)

28. Plachy, J., Becvar, Z., Strinati, E. C.: Dynamic resource allocation exploiting mobility prediction in mobile edge computing. In: IEEE 27th Annual International Symposium on Personal, Indoor, and Mobile Radio Communications (PIMRC), pp. 1–6. Valencia (2016) https://doi.org/10.1109/pimrc.2016.7794955

29. Principali, L.: Rivoluzione smart mobility, così la "accende-ranno" 5G e IoT (2018). https://www.corrierecomunicazioni.it/digital-eco-nomy/rivoluzione-smart-mobility-cosi-la-accenderanno-5g-e-iot/ (retrieved 2 March, 2019)

30. Rose, D.: Smarter cities with edge computing (2018). https://www.smartcitiesworld.net/opinions/opinions/smarter-cities-with-edge-computing (retrieved 11 March 2019)

31. Sawat, S.: The Increasing Significance of Edge Analytics in Smart City Projects (2018). https://www.einfochips.com/blog/the-increas-ing-significance-of-edge-analytics-in-smart-city-projects/

32. Scekic, O., Nastic, S., Dustdar, S.: Blockchain-supported smart city platform for social value co-creation and exchange. IEEE Int Comput **23**(1), 19–28 (2019). https://doi.org/10.1109/MIC.2018.2881518

33. Schaffers, H., Komninos, N., Pallot, M., Trousse, B., Nilsson, M., and Oliveira, A.: Smart cities and the future internet: towards cooperation frameworks for open innovation. In The future internet assembly, pp. 431–446. Springer, Berlin, Heidelberg (2011)

34. Sharma, P.K., Park, Y.H.: Blockchain-based hybrid network architecture for the smart city. Fut. Gener. Comput. Syst. **86**, 650–655 (2018). https://doi.org/10.1016/j.fu-ture.2018.04.060

35. Sun, J., Yan, J., Zhang, K.Z.K.: Blockchain-based sharing services: What blockchain technology can contribute to smart cities. In: Financial Innovation 2016, (2) 26, Springer. https://doi.org/10.1186/s40854-016-0040-y

36. Townsend, A.M.: Smart cities: Big data, civic hackers, and the quest for a new utopia. WW Norton & Company. (2013)

37. Trends (ITT), Dubai, United Arab Emirates, pp. 213–218 (2018). https://doi.org/10.1109/ctit.2018.8649542

38. Voda, A.I.; Radu, L.D.: Artificial intelligence and the future of smart cities. BRAIN. Broad Res. Artif. Intell. Neurosci. [S.l.] **9**(2), 110–127 (2018)

Printed in the United States
By Bookmasters